The Age Of Invention
A Chronicle Of Mechanical Conquest

by

Holland Thompson

The Age Of Invention
A Chronicle Of Mechanical Conquest
by Holland Thompson

ISBN: 978-93-59327-50-1

Published by

DOUBLE 9 BOOKS

2/13-B, Ansari Road
Daryaganj, New Delhi – 110002
info@double9books.com
www.double9books.com
Tel. 011-40042856

ABOUT THE AUTHOR

Holland McTyeire Thompson (July 30, 1873 – October 21, 1940) was a New South historian and author. Thompson was born in North Carolina's Randolph County. He got a bachelor's degree from the University of North Carolina. Thompson taught at Concord High School in Concord, North Carolina, from 1895 to 1899, and while there, he authored an essay about the transformation of southern culture from a rural agricultural to a textile/manufacturing way of life that he witnessed. This thesis helped Thompson get admission to Columbia University, where he earned his Ph.D. in 1901 and went on to become a full professor of history at City College of New York.

CONTENTS

CHAPTER I
BENJAMIN FRANKLIN AND HIS TIMES

On Milk Street, in Boston, opposite the Old South Church, lived Josiah Franklin, a maker of soap and candles. He had come to Boston with his wife about the year 1682 from the parish of Ecton, Northamptonshire, England, where his family had lived on a small freehold for about three hundred years. His English wife had died, leaving him seven children, and he had married a colonial girl, Abiah Folger, whose father, Peter Folger, was a man of some note in early Massachusetts.

Josiah Franklin was fifty-one and his wife Abiah thirty-nine, when the first illustrious American inventor was born in their house on Milk Street, January 17, 1706. He was their eighth child and Josiah's tenth son and was baptized Benjamin. What little we know of Benjamin's childhood is contained in his "Autobiography", which the world has accepted as one of its best books and which was the first American book to be so accepted. In the crowded household, where thirteen children grew to manhood and womanhood, there were no luxuries. Benjamin's period of formal schooling was less than two years, though he could never remember the time when he could not read, and at the age of ten he was put to work in his father's shop.

Benjamin was restless and unhappy in the shop. He appeared to have no aptitude at all for the business of soap making. His parents debated whether they might not educate him for the ministry, and his father took him into various shops in Boston, where he might see artisans at work, in the hope that he would be attracted to some trade. But Benjamin saw nothing there that he wished to engage in. He was inclined to follow the sea, as one of his older brothers had done.

His fondness for books finally determined his career. His older brother James was a printer, and in those days a printer was a literary man as well as a mechanic. The editor of a newspaper was always a printer and often composed his articles as he set them in type; so "composing" came to mean typesetting, and one who sets type is a compositor. Now James needed an apprentice. It happened then that young Benjamin, at the age of thirteen, was bound over by law to serve his brother.

James Franklin printed the "New England Courant", the fourth newspaper to be established in the colonies. Benjamin soon began to write articles for this newspaper. Then when his brother was put in jail, because he had printed matter considered libelous, and forbidden to continue as the publisher, the newspaper appeared in Benjamin's name.

The young apprentice felt that his brother was unduly severe and, after serving for about two years, made up his mind to run away. Secretly he took passage on a sloop and in three days reached New York, there to find that the one printer in the town, William Bradford, could give him no work. Benjamin then set out for Philadelphia. By boat to Perth Amboy, on foot to Burlington, and then by boat to Philadelphia was the course of his journey, which consumed five days. On a Sunday morning in October, 1723, the tired, hungry boy landed upon the Market Street wharf, and at once set out to find food and explore America's metropolis.

Benjamin found employment with Samuel Keimer, an eccentric printer just beginning business, and lodgings at the house of Read, whose daughter Deborah was later to become his wife. The intelligent young printer soon attracted the notice of Sir William Keith, Governor of Pennsylvania, who promised to set him up in business. First, however, he must go to London to buy a printing outfit. On the Governor's promise to send a letter of credit for his needs in London, Franklin set sail; but the Governor broke his word, and Franklin was obliged to remain in London nearly two years working at his trade. It was in London that he printed the first of his many pamphlets, an attack on revealed religion, called "A Dissertation on Liberty and Necessity, Pleasure and Pain." Though he met some interesting persons, from each of whom he extracted, according to his custom, every particle of information possible, no future opened for him in London, and he accepted an offer to return to Philadelphia with employment as a clerk. But early in 1727 his employer died, and Benjamin went back to his trade, as printers always do. He found work again in Keimer's printing office. Here his mechanical ingenuity and general ability presently began to appear; he invented a method of casting type, made ink, and became, in fact, the real manager of the business.

The ability to make friends was one of Franklin's traits, and the number of his acquaintances grew rapidly, both in Pennsylvania and New Jersey. "I grew convinced," he naively says, "that TRUTH, SINCERITY, and INTEGRITY in dealings between man and man were of the utmost importance to the felicity of life." Not long after his return from England he founded in Philadelphia the Junto, a society which at its regular meetings argued various questions and criticized the writings of the members. Through this society he enlarged his reputation as well as his education.

The father of an apprentice at Keimer's furnished the money to buy a printing outfit for his son and Franklin, but the son soon sold his share, and Benjamin Franklin, Printer, was fairly established in business at the age of twenty-four. The writing of an anonymous pamphlet on "The Nature and Necessity of a Paper Currency" called attention to the need of a further issue of paper money in Pennsylvania, and the author of the tract was rewarded with the contract to print the money, "a very profitable job, and a great help to me." Small favors were thankfully received. And, "I took care not only to be in REALITY industrious and frugal, but to avoid all appearances to the contrary. I drest plainly; I was seen at no places of idle diversion." And, "to show that I was not above my business, I sometimes brought home the paper I purchased at the stores thru the streets on a wheelbarrow."

"The Universal Instructor in All Arts and Sciences and Pennsylvania Gazette": this was the high-sounding name of a newspaper which Franklin's old employer, Keimer, had started in Philadelphia. But bankruptcy shortly overtook Keimer, and Franklin took the newspaper with its ninety subscribers. The "Universal Instructor" feature of the paper consisted of a page or two weekly of "Chambers's Encyclopedia". Franklin eliminated this feature and dropped the first part of the long name. "The Pennsylvania Gazette" in Franklin's hands soon became profitable. And it lives today in the fullness of abounding life, though under another name. "Founded A.D. 1728 by Benj. Franklin" is the proud legend of "The Saturday Evening Post", which carries on, in our own times, the Franklin tradition.

The "Gazette" printed bits of local news, extracts from the London "Spectator", jokes, verses, humorous attacks on Bradford's "Mercury", a rival paper, moral essays by the editor, elaborate hoaxes, and pungent political or social criticism. Often the editor wrote and printed letters to himself, either to emphasize some truth or to give him the opportunity to ridicule some folly in a reply to "Alice Addertongue," "Anthony Afterwit," or other mythical but none the less typical person.

If the countryman did not read a newspaper, or buy books, he was, at any rate, sure to own an almanac. So in 1732 Franklin brought out "Poor Richard's Almanac". Three editions were sold within a few months. Year after year the sayings of Richard Saunders, the alleged publisher, and Bridget, his wife, creations of Franklin's fancy, were printed in the almanac. Years later the most striking of these sayings were collected and published. This work has been translated into as many as twenty languages and is still in circulation today.

Franklin kept a shop in connection with his printing office, where he sold a strange variety of goods: legal blanks, ink, pens, paper, books, maps, pictures, chocolate, coffee, cheese, codfish, soap, linseed oil, broadcloth, Godfrey's cordial, tea, spectacles, rattlesnake root, lottery tickets, and stoves—to mention only a few of the many articles he advertised. Deborah Read, who became his wife in 1730, looked after his house, tended shop, folded and stitched pamphlets, bought rags, and helped him to live economically. "We kept no idle servants," says Franklin, "our table was plain and simple, our furniture of the cheapest. For instance, my breakfast was a long time bread and milk (no tea), and I ate it out of a twopenny earthen porringer with a pewter spoon."

With all this frugality, Franklin was not a miser; he abhorred the waste of money, not the proper use. His wealth increased rapidly. "I experienced too," he says, "the truth of the observation, 'THAT AFTER GETTING THE FIRST HUNDRED POUND, IT IS MORE EASY TO GET THE SECOND, money itself being of a prolific nature." He gave much unpaid public service and subscribed generously to public purposes; yet he was able, at the early age of forty-two, to turn over his printing office to one of his journeymen, and to retire from active business, intending to devote himself thereafter to such public employment as should come his way, to philosophical or scientific studies, and to amusements.

From boyhood Franklin had been interested in natural phenomena. His "Journal of a Voyage from London to Philadelphia", written at sea as he returned from his first stay in London, shows unusual powers of exact observation for a youth of twenty. Many of the questions he propounded to the Junto had a scientific bearing. He made an original and important invention in 1749, the "Pennsylvania fireplace," which, under the name of the Franklin stove, is in common use to this day, and which brought to the ill-made houses of the time increased comfort and a great saving of fuel. But it brought Franklin no pecuniary reward, for he never deigned to patent any of his inventions.

His active, inquiring mind played upon hundreds of questions in a dozen different branches of science. He studied smoky chimneys; he invented bifocal spectacles; he studied the effect of oil upon ruffled water; he identified the "dry bellyache" as lead poisoning; he preached ventilation in the days when windows were closed tight at night, and upon the sick at all times; he investigated fertilizers in agriculture. Many of his suggestions have since borne fruit, and his observations show that he foresaw some of the great developments of the nineteenth century.

His fame in science rests chiefly upon his discoveries in electricity. On a visit to Boston in 1746 he saw some electrical experiments and at once became deeply interested. Peter Collinson of London, a Fellow of the Royal

Society, who had made several gifts to the Philadelphia Library, sent over some of the crude electrical apparatus of the day, which Franklin used, as well as some contrivances he had purchased in Boston. He says in a letter to Collinson: "For my own part, I never was before engaged in any study that so engrossed my attention and my time as this has lately done."

Franklin's letters to Collinson tell of his first experiments and speculations as to the nature of electricity. Experiments made by a little group of friends showed the effect of pointed bodies in drawing off electricity. He decided that electricity was not the result of friction, but that the mysterious force was diffused through most substances, and that nature is always alert to restore its equilibrium. He developed the theory of positive and negative electricity, or plus and minus electrification. The same letter tells of some of the tricks which the little group of experimenters were accustomed to play upon their wondering neighbors. They set alcohol on fire, relighted candles just blown out, produced mimic flashes of lightning, gave shocks on touching or kissing, and caused an artificial spider to move mysteriously.

Franklin carried on experiments with the Leyden jar, made an electrical battery, killed a fowl and roasted it upon a spit turned by electricity, sent a current through water and found it still able to ignite alcohol, ignited gunpowder, and charged glasses of wine so that the drinkers received shocks. More important, perhaps, he began to develop the theory of the identity of lightning and electricity, and the possibility of protecting buildings by iron rods. By means of an iron rod he brought down electricity into his house, where he studied its effect upon bells and concluded that clouds were generally negatively electrified. In June, 1752, he performed the famous experiment with the kite, drawing down electricity from the clouds and charging a Leyden jar from the key at the end of the string.

Franklin's letters to Collinson were read before the Royal Society but were unnoticed. Collinson gathered them together, and they were published in a pamphlet which attracted wide attention. Translated into French, they created great excitement, and Franklin's conclusions were generally accepted by the scientific men of Europe. The Royal Society, tardily awakened, elected Franklin a member and in 1753 awarded him the Copley medal with a complimentary address.*

* It may be useful to mention some of the scientific facts and mechanical principles which were known to Europeans at this time. More than one learned essay has been written to prove the mechanical indebtedness of the modern world to the ancient, particularly to the works of those mechanically minded Greeks: Archimedes, Aristotle, Ctesibius, and Hero of Alexandria. The Greeks employed the lever, the tackle, and the crane, the force-pump, and

the suction-pump. They had discovered that steam could be mechanically applied, though they never made any practical use of steam. In common with other ancients they knew the principle of the mariner's compass. The Egyptians had the water-wheel and the rudimentary blast-furnace. The pendulum clock appears to have been an invention of the Middle Ages. The art of printing from movable type, beginning with Gutenberg about 1450, helped to further the Renaissance. The improved mariner's compass enabled Columbus to find the New world; gunpowder made possible its conquest. The compound microscope and the first practical telescope came from the spectacle makers of Middelburg, Holland, the former about 1590 and the latter about 1608. Harvey, an English physician, had discovered the circulation of the blood in 1628, and Newton, an English mathematician, the law of gravitation in 1685.

If Franklin's desire to continue his scientific researches had been gratified, it is possible that he might have discovered some of the secrets for which the world waited until Edison and his contemporaries revealed them more than a century later. Franklin's scientific reputation has grown with the years, and some of his views seem in perfect accord with the latest developments in electricity. But he was not to be permitted to continue his experiments. He had shown his ability to manage men and was to be called to a wider field.

Franklin's influence among his fellow citizens in Philadelphia was very great. Always ostensibly keeping himself in the background and working through others, never contradicting, but carrying his point by shrewd questions which showed the folly of the contrary position, he continued to set on foot and carry out movements for the public good. He established the first circulating library in Philadelphia, and one of the first in the country, and an academy which grew into the University of Pennsylvania. He was instrumental in the foundation of a hospital. "I am often ask'd by those to whom I propose subscribing," said one of the doctors who had made fruitless attempts to raise money for the hospital, "Have you consulted Franklin upon this business?" Other public matters in which the busy printer was engaged were the paving and cleaning of the streets, better street lighting, the organization of a police force and of a fire company. A pamphlet which he published, "Plain Truth", showing the helplessness of the colony against the French and Indians, led to the organization of a volunteer militia, and funds were raised for arms by a lottery. Franklin himself was elected colonel of the Philadelphia regiment, "but considering myself unfit, I declined the station and recommended Mr. Lawrence, a fine person and man of influence, who was accordingly appointed." In spite of his militarism, Franklin retained the position which he held as Clerk of the

Assembly, though the majority of the members were Quakers opposed to war on principle.

The American Philosophical Society owes its origin to Franklin. It was formally organized on his motion in 1743, but the society has accepted the organization of the Junto in 1727 as the actual date of its birth. From the beginning the society has had among its members many leading men of scientific attainments or tastes, not only of Philadelphia, but of the world. In 1769 the original society was consolidated with another of similar aims, and Franklin, who was the first secretary of the society, was elected president and served until his death. The first important undertaking was the successful observation of the transit of Venus in 1769, and many important scientific discoveries have since been made by its members and first given to the world at its meetings.

Franklin's appointment as one of the two Deputy Postmasters General of the colonies in 1753 enlarged his experience and his reputation. He visited nearly all the post offices in the colonies and introduced many improvements into the service. In none of his positions did his transcendent business ability show to better advantage. He established new postal routes and shortened others. There were no good roads in the colonies, but his post riders made what then seemed wonderful speed. The bags were opened to newspapers, the carrying of which had previously been a private and unlawful perquisite of the riders. Previously there had been one mail a week in summer between New York and Philadelphia and one a month in winter. The service was increased to three a week in summer and one in winter.

The main post road ran from northern New England to Savannah, closely hugging the seacoast for the greater part of the way. Some of the milestones set by Franklin to enable the postmasters to compute the postage, which was fixed according to distance, are still standing. Crossroads connected some of the larger communities away from the seacoast with the main road, but when Franklin died, after serving also as Postmaster General of the United States, there were only seventy-five post offices in the entire country.

Franklin took a hand in the final struggle between France and England in America. On the eve of the conflict, in 1754, commissioners from the several colonies were ordered to convene at Albany for a conference with the Six Nations of the Iroquois, and Franklin was one of the deputies from Pennsylvania. On his way to Albany he "projected and drew a plan for the union of all the colonies under one government so far as might be necessary for defense and other important general purposes." This statesmanlike "Albany Plan of Union," however, came to nothing. "Its fate was singular," says Franklin; "the assemblies did not adopt it, as they all thought there was too much PREROGATIVE in it and in England it was judg'd to have too much of the DEMOCRATIC."

How to raise funds for defense was always a grave problem in the colonies, for the assemblies controlled the purse-strings and released them with a grudging hand. In face of the French menace, this was Governor Shirley's problem in Massachusetts, Governor Dinwiddie's in Virginia, and Franklin's in the Quaker and proprietary province of Pennsylvania. Franklin opposed Shirley's suggestion of a general tax to be levied on the colonies by Parliament, on the ground of no taxation without representation, but used all his arts to bring the Quaker Assembly to vote money for defense, and succeeded. When General Braddock arrived in Virginia Franklin was sent by the Assembly to confer with him in the hope of allaying any prejudice against Quakers that the general might have conceived. If that blustering and dull-witted soldier had any such prejudice, it melted away when the envoy of the Quakers promised to procure wagons for the army. The story of Braddock's disaster does not belong here, but Franklin formed a shrewd estimate of the man which proved accurate. His account of Braddock's opinion of the colonial militia is given in a sentence: "He smil'd at my ignorance, and reply'd, 'These savages may, indeed, be a formidable enemy to your raw American militia, but upon the King's regular and disciplin'd troops, sir, it is impossible they should make any impression.'" After Braddock's defeat the Pennsylvania Assembly voted more money for defense, and the unmilitary Franklin was placed in command of the frontier with full power. He built forts, as he had planned, and incidentally learned much of the beliefs of a group of settlers in the back country, the "Unitas Fratrum," better known as the Moravians.

The death struggle between English and French in America served only to intensify a lesser conflict that was being waged between the Assembly and the proprietors of Pennsylvania; and the Assembly determined to send Franklin to London to seek judgment against the proprietors and to request the King to take away from them the government of Pennsylvania. Franklin, accompanied by his son William, reached London in July, 1757, and from this time on his life was to be closely linked with Europe. He returned to America six years later and made a trip of sixteen hundred miles inspecting postal affairs, but in 1764 he was again sent to England to renew the petition for a royal government for Pennsylvania, which had not yet been granted. Presently that petition was made obsolete by the Stamp Act, and Franklin became the representative of the American colonies against King and Parliament.

Franklin did his best to avert the Revolution. He made many friends in England, wrote pamphlets and articles, told comical stories and fables where they might do some good, and constantly strove to enlighten the ruling class of England upon conditions and sentiment in the colonies. His examination before the House of Commons in February, 1766, marks perhaps the zenith of his intellectual powers. His wide knowledge, his wonderful poise, his ready wit, his marvelous gift for clear and epigrammatic statement, were never exhibited to better advantage and no doubt hastened the repeal of the Stamp Act. Franklin remained in England nine years longer, but his efforts to reconcile the conflicting claims of Parliament and the colonies were of no avail, and early in 1775 he sailed for home.

Franklin's stay in America lasted only eighteen months, yet during that time he sat in the Continental Congress and as a member of the most important committees; submitted a plan for a union of the colonies; served as Postmaster General and as chairman of the Pennsylvania Committee of Safety; visited Washington at Cambridge; went to Montreal to do what he could for the cause of independence in Canada; presided over the convention which framed a constitution for Pennsylvania; was a member of the committee appointed to draft the Declaration of Independence and of the committee sent on the futile mission to New York to discuss terms of peace with Lord Howe.

In September, 1776, Franklin was appointed envoy to France and sailed soon afterwards. The envoys appointed to act with him proved a handicap rather than a help, and the great burden of a difficult and momentous mission was thus laid upon an old man of seventy. But no other American could have taken his place. His reputation in France was already made, through his books and inventions and discoveries. To the corrupt and licentious court he was the personification of the age of simplicity, which it was the fashion to admire; to the learned, he was a sage; to the common man he was the apotheosis of all the virtues; to the rabble he was little less than a god. Great ladies sought his smiles; nobles treasured a kindly word; the shopkeeper hung his portrait on the wall; and the people drew aside in the streets that he might pass without annoyance. Through all this adulation Franklin passed serenely, if not unconsciously.

The French ministers were not at first willing to make a treaty of alliance, but under Franklin's influence they lent money to the struggling colonies. Congress sought to finance the war by the issue of paper currency and by borrowing rather than by taxation, and sent bill after bill to Franklin, who somehow managed to meet them by putting his pride in his pocket, and applying again and again to the French Government. He fitted out privateers and negotiated with the British concerning prisoners. At length he won from France recognition of the United States and then the Treaty of Alliance.

Not until two years after the Peace of 1783 would Congress permit the veteran to come home. And when he did return in 1785 his people would not allow him to rest. At once he was elected President of the Council of Pennsylvania and twice reelected in spite of his protests. He was sent to the Convention of 1787 which framed the Constitution of the United States. There he spoke seldom but always to the point, and the Constitution is the better for his suggestions. With pride he axed his signature to that great instrument, as he had previously signed the Albany Plan of Union, the Declaration of Independence, and the Treaty of Paris.

Benjamin Franklin's work was done. He was now an old man of eighty-two summers and his feeble body was racked by a painful malady. Yet he kept his face towards the morning. About a hundred of his letters, written after this time, have been preserved. These letters show no retrospection, no looking backward. They never mention "the good old times." As long as he lived, Franklin looked forward. His interest in the mechanical arts and in scientific progress seems never to have abated. He writes in October, 1787, to a friend in France, describing his experience with lightning conductors and referring to the work of David Rittenhouse, the celebrated astronomer of Philadelphia. On the 31st of May in the following year he is writing to the Reverend John Lathrop of Boston:

"I have long been impressed with the same sentiments you so well express, of the growing felicity of mankind, from the improvement in philosophy, morals, politics, and even the conveniences of common living, and the invention of new and useful utensils and instruments; so that I have sometimes wished it had been my destiny to be born two or three centuries hence. For invention and improvement are prolific, and beget more of their kind. The present progress is rapid. Many of great importance, now unthought of, will, before that period, be produced."

Thus the old philosopher felt the thrill of dawn and knew that the day of great mechanical inventions was at hand. He had read the meaning of the puffing of the young steam engine of James Watt and he had heard of a marvelous series of British inventions for spinning and weaving. He saw that his own countrymen were astir, trying to substitute the power of steam for the strength of muscles and the fitful wind. John Fitch on the Delaware and James Rumsey on the Potomac were already moving vessels by steam. John Stevens of New York and Hoboken had set up a machine shop that was to mean much to mechanical progress in America. Oliver Evans, a mechanical genius of Delaware, was dreaming of the application of high-pressure steam to both road and water carriages. Such manifestations, though still very faint, were to Franklin the signs of a new era.

And so, with vision undimmed, America's most famous citizen lived on until near the end of the first year of George Washington's administration. On April 17, 1790, his unconquerable spirit took its flight.

In that year, 1790, was taken the First Census of the United States. The new nation had a population of about four million people. It then included practically the present territory east of the Mississippi, except the Floridas, which belonged to Spain. But only a small part of this territory was occupied. Much of New York and Pennsylvania was savage wilderness. Only the seacoast of Maine was inhabited, and the eighty-two thousand inhabitants of Georgia hugged the Savannah River. Hardy pioneers had climbed the Alleghanies into Kentucky and Tennessee, but the Northwest Territory—comprising Ohio, Michigan, Indiana, Illinois, and Wisconsin— was not enumerated at all, so scanty were its people, perhaps not more than four thousand.

Though the First Census did not classify the population by occupation it is certain that nine-tenths of the breadwinners worked more or less upon the soil. The remaining tenth were engaged in trade, transportation, manufacturing, fishing and included also the professional men, doctors, lawyers, clergymen, teachers, and the like. In other words, nine out of ten of the population were engaged primarily in the production of food, an occupation which today engages less than three out of ten. This comparison, however, requires some qualification. The farmer and the farmer's wife and children performed many tasks which are now done in factories. The successful farmer on the frontier had to be a jack of many trades. Often he tanned leather and made shoes for his family and harness for his horses. He was carpenter, blacksmith, cobbler, and often boat-builder and fisherman as well. His wife made soap and candles, spun yarn and dyed it, wove cloth and made the clothes the family wore, to mention only a few of the tasks of the women of the eighteenth century.

The organization of industry, however, was beginning. Here and there were small paper mills, glass factories-though many houses in the back country were without glass windows—potteries, and iron foundries and forges. Capitalists, in some places, had brought together a few handloom weavers to make cloth for sale, and the famous shoemakers of Massachusetts commonly worked in groups.

The mineral resources of the United States were practically unknown. The country seems to have produced iron enough for its simple needs, some coal, copper, lead, gold, silver, and sulphur. But we may say that mining was hardly practiced at all.

The fisheries and the shipyards were great sources of wealth, especially for New England. The cod fishers numbered several hundred vessels and the whalers about forty. Thousands of citizens living along the seashore and the rivers fished more or less to add to the local food supply. The deep-sea fishermen exported a part of their catch, dried and salted. Yankee vessels sailed to all ports of the world and carried the greater part of the foreign commerce of the United States. Flour, tobacco, rice, wheat, corn, dried fish, potash, indigo, and staves were the principal exports. Great Britain was the best customer, with the French West Indies next, and then the British West Indies. The principal imports came from the same countries. Imports and exports practically balanced each other, at about twenty million dollars annually, or about five dollars a head. The great merchants owned ships and many of them, such as John Hancock of Boston, and Stephen Girard of Philadelphia, had grown very rich.

Inland transportation depended on horses and oxen or boats. There were few good roads, sometimes none at all save bridle paths and trails. The settlers along the river valleys used boats almost entirely. Stage-coaches made the journey from New York to Boston in four days in summer and in six in winter. Two days were required to go between New York and Philadelphia. Forty to fifty miles a day was the speed of the best coaches, provided always that they did not tumble into the ditch. In many parts of the country one must needs travel on horseback or on foot.

Even the wealthiest Americans of those days had few or none of the articles which we regard today as necessities of life. The houses were provided with open—which, however cheerful, did not keep them warm— or else with Franklin's stoves. To strike a fire one must have the flint and tinderbox, for matches were unknown until about 1830. Candles made the darkness visible. There was neither plumbing nor running water. Food was cooked in the ashes or over an open fire.

The farmer's tools were no less crude than his wife's. His plough had been little improved since the days of Rameses. He sowed his wheat by hand, cut it with a sickle, flailed it out upon the floor, and laboriously winnowed away the chaff.

In that same year, 1790, came a great boon and encouragement to inventors, the first Federal Patent Act, passed by Congress on the 10th of April. Every State had its own separate patent laws or regulations, as an inheritance from colonial days, but the Fathers of the Constitution had wisely provided that this function of government should be exercised by the nation.* The Patent Act, however, was for a time unpopular, and some

States granted monopolies, particularly of transportation, until they were forbidden to do so by judicial decision.

The first Patent Act provided that an examining board, consisting of the Secretary of State, the Secretary of War, and the Attorney-General, or any two of them, might grant a patent for fourteen years, if they deemed the invention useful and important. The patent itself was to be engrossed and signed by the President, the Secretary of State, and the Attorney-General. And the cost was to be three dollars and seventy cents, plus the cost of copying the specifications at ten cents a sheet.

The first inventor to avail himself of the advantages of the new Patent Act was Samuel Hopkins of Vermont, who received a patent on the 31st of July for an improved method of "Making Pot and Pearl Ashes." The world knows nothing of this Samuel Hopkins, but the potash industry, which was evidently on his mind, was quite important in his day. Potash, that is, crude potassium carbonate, useful in making soap and in the manufacture of glass, was made by leaching wood ashes and boiling down the lye. To produce a ton of potash, the trees on an acre of ground would be cut down and burned, the ashes leached, and the lye evaporated in great iron kettles. A ton of potash was worth about twenty-five dollars. Nothing could show more plainly the relative value of money and human labor in those early times.

Two more patents were issued during the year 1790. The second went to Joseph S. Sampson of Boston for a method of making candles, and the third to Oliver Evans, of whom we shall learn more presently, for an improvement in manufacturing flour and meal. The fourth patent was granted in 1791 to Francis Baily of Philadelphia for making punches for types. Next Aaron Putnam of Medford, Massachusetts, thought that he could improve methods of distilling, and John Stone of Concord, Massachusetts, offered a new method of driving piles for bridges. And a versatile inventor, Samuel Mulliken of Philadelphia, received four patents in one day for threshing grain, cutting and polishing marble, raising a nap on cloth, and breaking hemp.

Then came improvements in making nails, in making bedsteads, in the manufacture of boats, and for propelling boats by cattle. On August 26, 1791, James Rumsey, John Stevens, and John Fitch (all three will appear again in this narrative) took out patents on means of propelling boats. On the same day Nathan Read received one on a process for distilling alcohol.

More than fifty patents were granted under the Patent Act of 1790, and mechanical devices were coming in so thick and fast that the department heads apparently found it inconvenient to hear applications. So the Act of 1790 was repealed. The second Patent Act (1793) provided that a patent should be granted as a matter of routine to any one who swore to the originality of his device and paid the sum of thirty dollars as a fee. No one except a citizen, however, could receive a patent. This act, with some amendments, remained in force until 1836, when the present Patent Office was organized with a rigorous and intricate system for examination of all claims in order to prevent interference. Protection of the property rights of inventors has been from the beginning of the nation a definite American policy, and to this policy may be ascribed innumerable inventions which have contributed to the greatness of American industry and multiplied the world's comforts and conveniences.

Under the second Patent Act came the most important invention yet offered, an invention which was to affect generations then unborn. This was a machine for cleaning cotton and it was offered by a young Yankee schoolmaster, temporarily sojourning in the South.

CHAPTER II
ELI WHITNEY AND THE COTTON GIN

The cotton industry is one of the most ancient. One or more of the many species of the cotton plant is indigenous to four continents, Asia, Africa, and the Americas, and the manufacture of the fiber into yarn and cloth seems to have developed independently in each of them. We find mention of cotton in India fifteen hundred years before Christ. The East Indians, with only the crudest machinery, spun yarn and wove cloth as diaphanous as the best appliances of the present day have been able to produce.

Alexander the Great introduced the "vegetable wool" into Europe. The fable of the "vegetable lamb of Tartary" persisted almost down to modern times. The Moors cultivated cotton in Spain on an extensive scale, but after their expulsion the industry languished. The East India Company imported cotton fabrics into England early in the seventeenth century, and these fabrics made their way in spite of the bitter opposition of the woolen interests, which were at times strong enough to have the use of cotton cloth prohibited by law. But when the Manchester spinners took up the manufacture of cotton, the fight was won. The Manchester spinners, however, used linen for their warp threads, for without machinery they could not spin threads sufficiently strong from the short-fibered Indian cotton.

In the New World the Spanish explorers found cotton and cotton fabrics in use everywhere. Columbus, Cortes, Pizarro, Magellan, and others speak of the various uses to which the fiber was put, and admired the striped awnings and the colored mantles made by the natives. It seems probable that cotton was in use in the New World quite as early as in India.

The first English settlers in America found little or no cotton among the natives. But they soon began to import the fiber from the West Indies, whence came also the plant itself into the congenial soil and climate of the Southern colonies. During the colonial period, however, cotton never became the leading crop, hardly an important crop. Cotton could be grown profitably only where there was an abundant supply of exceedingly cheap labor, and labor in America, white or black, was never and could never be as cheap as in India. American slaves could be much more profitably employed in the cultivation of rice and indigo.

Three varieties of the cotton plant were grown in the South. Two kinds of the black-seed or long-staple variety thrived in the sea-islands and along the coast from Delaware to Georgia, but only the hardier and more prolific green-seed or short-staple cotton could be raised inland. The labor of cultivating and harvesting cotton of any kind was very great. The fiber, growing in bolls resembling a walnut in size and shape, had to be taken by hand from every boll, as it has to be today, for no satisfactory cotton harvester has yet been invented. But in the case of the green-seed or upland cotton, the only kind which could ever be cultivated extensively in the South, there was another and more serious obstacle in the way, namely, the difficulty of separating the fiber from the seeds. No machine yet devised could perform this tedious and unprofitable task. For the black-seed or sea-island cotton, the churka, or roller gin, used in India from time immemorial, drawing the fiber slowly between a pair of rollers to push out the seeds, did the work imperfectly, but this churka was entirely useless for the green-seed variety, the fiber of which clung closely to the seed and would yield only to human hands. The quickest and most skillful pair of hands could separate only a pound or two of lint from its three pounds of seeds in an ordinary working day. Usually the task was taken up at the end of the day, when the other work was done. The slaves sat round an overseer who shook the dozing and nudged the slow. It was also the regular task for a rainy day. It is not surprising, then, that cotton was scarce, that flax and wool in that day were the usual textiles, that in 1783 wool furnished about seventy-seven per cent, flax about eighteen per cent, and cotton only about five per cent of the clothing of the people of Europe and the United States.

That series of inventions designed for the manufacture of cloth, and destined to transform Great Britain, the whole world, in fact, was already completed in Franklin's time. Beginning with the flying shuttle of John Kay in 1738, followed by the spinning jenny of James Hargreaves in 1764, the water-frame of Richard Arkwright in 1769, and the mule of Samuel Crompton ten years later, machines were provided which could spin any quantity of fiber likely to be offered. And when, in 1787, Edmund Cartwright, clergyman and poet, invented the self-acting loom to which power might be applied, the series was complete. These inventions, supplementing the steam engine of James Watt, made the Industrial Revolution. They destroyed the system of cottage manufactures in England and gave birth to the great textile establishments of today.

The mechanism for the production of cloth on a great scale was provided, if only the raw material could be found.

The romance of cotton begins on a New England farm. It was on a farm in the town (township) of Westboro, in Worcester County, Massachusetts, in the year 1765, that Eli Whitney, inventor of the cotton gin, was born. Eli's father was a man of substance and standing in the community, a mechanic as well as a farmer, who occupied his leisure in making articles for his neighbors. We are told that young Eli displayed a passion for tools almost as soon as he could walk, that he made a violin at the age of twelve and about the same time took his father's watch to pieces surreptitiously and succeeded in putting it together again so successfully as to escape detection. He was able to make a table knife to match the others of a broken set. As a boy of fifteen or sixteen, during the War of Independence, he was supplying the neighborhood with hand-made nails and various other articles. Though he had not been a particularly apt pupil in the schools, he conceived the ambition of attending college; and so, after teaching several winters in rural schools, he went to Yale. He appears to have paid his own way through college by the exercise of his mechanical talents. He is said to have mended for the college some imported apparatus which otherwise would have had to go to the old country for repairs. "There was a good mechanic spoiled when you came to college," he was told by a carpenter in the town. There was no "Sheff" at Yale in those days to give young men like Whitney scientific instruction; so, defying the bent of his abilities, Eli went on with his academic studies, graduated in 1792, at the age of twenty-seven, and decided to be a teacher or perhaps a lawyer.

Like so many young New Englanders of the time, Whitney sought employment in the South. Having received the promise of a position in South Carolina, he embarked at New York, soon after his graduation, on a sailing vessel bound for Savannah. On board he met the widow of General Nathanael Greene of Revolutionary fame, and this lady invited him to visit her plantation at Mulberry Grove, near Savannah. What happened then is best told by Eli Whitney himself, in a letter to his father, written at New Haven, after his return from the South some months later, though the spelling master will probably send Whitney to the foot of the class:

"New Haven, Sept. 11th, 1793.

"... I went from N. York with the family of the late Major General Greene to Georgia. I went immediately with the family to their Plantation about twelve miles from Savannah with an expectation of spending four or five days and then proceed into Carolina to take the school as I have mentioned in former letters. During this time I heard much said of the extreme difficulty of ginning Cotton, that is, separating it from its seeds. There were a number of very respectable Gentlemen at Mrs. Greene's who all agreed that if a machine could be invented which would clean the cotton with expedition, it would be a great thing both to the Country and to the inventor. I involuntarily happened to be thinking on the subject and struck out a plan of a Machine in my mind, which I communicated to Miller (who is agent to the Executors of Genl. Greene and resides in the family, a man of respectability and property), he was pleased with the Plan and said if I would pursue it and try an experiment to see if it would answer, he would be at the whole expense, I should loose nothing but my time, and if I succeeded we would share the profits. Previous to this I found I was like to be disappointed in my school, that is, instead of a hundred, I found I could get only fifty Guineas a year. I however held the refusal of the school until I tried some experiments. In about ten Days I made a little model, for which I was offered, if I would give up all right and title to it, a Hundred Guineas. I concluded to relinquish my school and turn my attention to perfecting the Machine. I made one before I came away which required the labor of one man to turn it and with which one man will clean ten times as much cotton as he can in any other way before known and also cleanse it much better than in the usual mode. This machine may be turned by water or with a horse, with the greatest ease, and one man and a horse will do more than fifty men with the old machines. It makes the labor fifty times less, without throwing any class of People out of business.

"I returned to the Northward for the purpose of having a machine made on a large scale and obtaining a Patent for the invention. I went to Philadelphia* soon after I arrived, made myself acquainted with the steps necessary to obtain a Patent, took several of the steps and the Secretary of State Mr. Jefferson agreed to send the Patent to me as soon it could be made out—so that I apprehended no difficulty in obtaining the Patent—Since I have been here I have employed several workmen in making machines and as soon as my business is such that I can leave it a few days, I shall come to Westboro'**. I think it is probable I shall go to Philadelphia again before I come to Westboro', and when I do come I shall be able to stay but few days. I am certain I can obtain a patent in England. As soon as I have got a Patent in America I shall go with the machine which I am now making, to Georgia, where I shall stay a few weeks to see it at work. From thence I expect to go to England, where I shall probably continue two or three years. How advantageous this business will eventually prove to me, I cannot say. It is generally said by those who know anything about it, that I shall make a Fortune by it. I have no expectation that I shall make an independent fortune by it, but think I had better pursue it than any other business into which I can enter. Something which cannot be foreseen may frustrate my expectations and defeat my Plan; but I am now so sure of success that ten thousand dollars, if I saw the money counted out to me, would not tempt me to give up my right and relinquish the object. I wish you, sir, not to show this letter nor communicate anything of its contents to any body except My Brothers and Sister, ENJOINING it on them to keep the whole A PROFOUND SECRET."

* Then the national capital.

** Hammond, "Correspondence of Eli Whitney," American Historical Review, vol. III, p. 99. The other citations in this chapter are from the same source, unless otherwise stated.

The invention, however, could not be kept "a profound secret," for knowledge of it was already out in the cotton country. Whitney's hostess, Mrs. Greene, had shown the wonderful machine to some friends, who soon spread the glad tidings, and planters, near and far, had come to Mulberry Grove to see it. The machine was of very simple construction; any blacksmith or wheelwright, knowing the principle of the design, could make one. Even before Whitney could obtain his patent, cotton gins based on his were being manufactured and used.

Whitney received his patent in March, 1794, and entered on his new work with enthusiasm. His partner, Phineas Miller, was a cultivated New England gentleman, a graduate of Yale College, who, like Whitney, had sought his fortune as a teacher in the South. He had been a tutor in the Greene household and on General Greene's death had taken over the management of his estates. He afterwards married Mrs. Greene. The partners decided to manufacture the machines in New Haven, Whitney to give his time to the production, Miller to furnish the capital and attend to the firm's interests in the South.

At the outset the partners blundered seriously in their plan for commercializing the invention. They planned to buy seed cotton and clean it themselves; also to clean cotton for the planters on the familiar toll system, as in grinding grain, taking a toll of one pound of cotton out of every three. "Whitney's plan in Georgia," says a recent writer, "as shown by his letters and other evidence, was to own all the gins and gin all the cotton made in the country. It is but human nature that this sort of monopoly should be odious to any community."* Miller appears to have calculated that the planters could afford to pay for the use of the new invention about one-half of all the profits they derived from its use. An equal division, between the owners of the invention on the one hand and the cotton growers on the other, of all the super-added wealth arising from the invention, seemed to him fair. Apparently the full meaning of such an arrangement did not enter his mind. Perhaps Miller and Whitney did not see at first that the new invention would cause a veritable industrial revolution, or that the system they planned, if it could be made effective, would make them absolute masters of the cotton country, with the most stupendous monopoly in the world. Nor do they appear to have realized that, considering the simple construction of their machine and the loose operation of the patent law at that time, the planters of the South would never submit to so great a tribute as they proposed to exact. Their attempt in the first instance to set up an unfair monopoly brought them presently into a sea of troubles, which they never passed out of, even when they afterwards changed their tack and offered to sell the machines with a license, or a license alone, at a reasonable price.

** Tompkins, "Cotton and Cotton Oil", p. 86.*

Misfortune pursued the partners from the beginning. Whitney writes to his father from New Haven in May, 1794, that his machines in Georgia are working well, but that he apprehends great difficulty in manufacturing them as fast as they are needed. In March of the following year he writes again, saying that his factory in New Haven has been destroyed by fire: "When I returned home from N. York I found my property all in ashes! My shop, all my tools, material and work equal to twenty finished cotton machines all gone. The manner in which it took fire is altogether unaccountable." Besides, the partners found themselves in distress for lack of capital. Then word came from England that the Manchester spinners had found the ginned cotton to contain knots, and this was sufficient to start the rumor throughout the South that Whitney's gin injured the cotton fiber and that cotton cleaned by them was worthless. It was two years before this ghost was laid. Meanwhile Whitney's patent was being infringed on every hand. "They continue to clean great quantities of cotton with Lyon's Gin and sell it advantageously while the Patent ginned cotton is run down as good for nothing," writes Miller to Whitney in September, 1797. Miller and Whitney brought suits against the infringers but they could obtain no redress in the courts.

Whitney's attitude of mind during these troubles is shown in his letters. He says the statement that his machines injure the cotton is false, that the source of the trouble is bad cotton, which he ventures to think is improved fifty per cent by the use of his gin, and that it is absurd to say that the cotton could be injured in any way in the process of cleaning. "I think," he says, writing to Miller, "you will be able to convince the CANDID that this is quite a mistaken notion and them that WILL NOT BELIEVE may be damn'd." Again, writing later to his friend Josiah Stebbins in New England: "I have a set of the most Depraved villains to combat and I might almost as well go

to HELL in search of HAPPINESS as apply to a Georgia Court for Justice."
And again: "You know I always believed in the 'DEPRAVITY OF HUMAN
NATURE.' I thought I was long ago sufficiently 'grounded and stablished'
in this Doctrine. But God Almighty is continually pouring down cataracts of
testimony upon me to convince me of this fact. 'Lord I believe, help thou,' not
'mine unbelief,' but me to overcome the rascality of mankind." His partner
Miller, on the other hand, is inclined to be more philosophical and suggests
to Whitney that "we take the affairs of this world patiently and that the
little dust which we may stir up about cotton may after all not make much
difference with our successors one hundred, much less one thousand years
hence." Miller, however, finally concluded that, "the prospect of making
anything by ginning in this State [Georgia] is at an end. Surreptitious gins
are being erected in every part of the country; and the jurymen at Augusta
have come to an understanding among themselves, that they will never give
a verdict in our favor, let the merits of the case be as they may."*

Cited in Roe, "English and American Tool Builders", p.

153.

Miller and Whitney were somewhat more fortunate in other States than
in Georgia though they nowhere received from the cotton gin enough to
compensate them for their time and trouble nor more than a pitiable fraction
of the great value of their invention. South Carolina, in 1801, voted them
fifty thousand dollars for their patent rights, twenty thousand dollars to be
paid down and the remainder in three annual payments of ten thousand
dollars each. "We get but a song for it," wrote Whitney, "in comparison
with the worth of the thing, but it is securing something." Why the partners
were willing to take so small a sum was later explained by Miller. They
valued the rights for South Carolina at two hundred thousand dollars,
but, since the patent law was being infringed with impunity, they were
willing to take half that amount; "and had flattered themselves," wrote
Miller, "that a sense of dignity and justice on the part of that honorable
body [the Legislature] would not have countenanced an offer of a less sum
than one hundred thousand dollars. Finding themselves, however, to be
mistaken in this opinion, and entertaining a belief that the failure of such
negotiation, after it commenced, would have a tendency to diminish the
prospect, already doubtful, of enforcing the Patent Law, it was concluded to
be best under existing circumstances to accept the very inadequate sum of
fifty thousand dollars offered by the Legislature and thereby relinquish and
entirely abandon three-fourths of the actual value of the property."

But even the fifty thousand dollars was not collected without difficulty.
South Carolina suspended the contract, after paying twenty thousand

dollars, and sued Miller and Whitney for recovery of the sum paid, on the ground that the partners had not complied with the conditions. Whitney succeeded, in 1805, in getting the Legislature to reinstate the contract and pay him the remainder of the money. Miller, discouraged and broken by the long struggle, had died in the meantime.

The following passage from a letter written by Whitney in February, 1805, to Josiah Stebbins, gives Whitney's views as to the treatment he had received at the hands of the authorities. He is writing from the residence of a friend near Orangeburg, South Carolina.

"The principal object of my present excursion to this Country was to get this business set right; which I have so far effected as to induce the Legislature of this State to recind all their former SUSPENDING LAWS and RESOLUTIONS, to agree once more to pay the sum of 30,000 Dollars which was due and make the necessary appropriations for that purpose. I have as yet however obtained but a small part of this payment. The residue is promised me in July next. Thus you see my RECOMPENSE OF REWARD is as the land of Canaan was to the Jews, resting a long while in promise. If the Nations with whom I have to contend are not as numerous as those opposed to the Israelites, they are certainly much greater HEATHENS, having their hearts hardened and their understanding blinded, to make, propagate and believe all manner of lies. Verily, Stebbins, I have had much vexation of spirit in this business. I shall spend forty thousand dollars to obtain thirty, and it will all end in vanity at last. A contract had been made with the State of Tennessee which now hangs SUSPENDED. Two attempts have been made to induce the State of No. Carolina to RECIND their CONTRACT, neither of which have succeeded. Thus you see Brother Steb. Sovreign and Independent States warped by INTEREST will be ROGUES and misled by Demagogues will be FOOLS. They have spent much time, MONEY and CREDIT, to avoid giving me a small compensation, for that which to them is worth millions."

Meanwhile North Carolina had agreed to buy the rights for the State on terms that yielded Whitney about thirty thousand dollars, and it is estimated that he received about ten thousand dollars from Tennessee, making his receipts in all about ninety thousand dollars, before deducting costs of litigation and other losses. The cotton gin was not profitable to its inventor. And yet no invention in history ever so suddenly transformed an industry and created enormous wealth. Eight years before Whitney's invention, eight bales of cotton, landed at Liverpool, were seized on the ground that so large a quantity of cotton could not have been produced in the United States. The year before that invention the United States exported less than one hundred and forty thousand pounds of cotton; the year after it, nearly half a million pounds; the next year over a million and a half; a year later still, over six million; by 1800, nearly eighteen million pounds a year. And by 1845 the United States was producing producing seven-eighths of the world's cotton. Today the United States produces six to eight billion pounds of cotton annually, and ninety-nine per cent of this is the upland or green-seed cotton, which is cleaned on the Whitney type of gin and was first made commercially available by Whitney's invention.*

** Roe, "English and American Tool Builders", pp. 150-51.*

More than half of this enormous crop is still exported in spite of the great demand at home. Cotton became and has continued to be the greatest single export of the United States. In ordinary years its value is greater than the combined value of the three next largest exports. It is on cotton that the United States has depended for the payment of its trade balance to Europe.

Other momentous results followed on the invention of the cotton gin. In 1793 slavery seemed a dying institution, North and South. Conditions of soil and climate made slavery unprofitable in the North. On many of the indigo, rice, and tobacco plantations in the South there were more slaves than could be profitably employed, and many planters were thinking of emancipating their slaves, when along came this simple but wonderful machine and with it the vision of great riches in cotton; for while slaves could not earn their keep separating the cotton from its seeds by hand, they could earn enormous profits in the fields, once the difficulty of extracting the seeds was solved. Slaves were no longer a liability but an asset. The price of "field hands" rose, and continued to rise. If the worn-out lands of the seaboard no longer afforded opportunity for profitable employment, the rich new lands of the Southwest called for laborers, and yet more laborers. Taking slaves with them, younger sons pushed out into the wilderness, became possessed of great tracts of fertile land, and built up larger plantations than those upon which they had been born. Cotton became King of the South.

The supposed economic necessity of slave labor led great men to defend slavery, and politics in the South became largely the defense of slavery against the aggression, real or fancied, of the free North. The rift between the sections became a chasm. Then came the War of Secession.

Though Miller was dead, Whitney carried on the fight for his rights in Georgia. His difficulties were increased by a patent which the Government at Philadelphia issued in May, 1796, to Hogden Holmes, a mechanic of Augusta, for an improvement in the cotton gin. The Holmes machines were soon in common use, and it was against the users of these that many of the suits for infringement were brought. Suit after suit ran its course in the Georgia courts, without a single decision in the inventor's favor. At length, however, in December, 1806, the validity of Whitney's patent was finally determined by decision of the United States Circuit Court in Georgia. Whitney asked for a perpetual injunction against the Holmes machine, and the court, finding that his invention was basic, granted him all that he asked.

By this time, however, the life of the patent had nearly run its course. Whitney applied to Congress for a renewal, but, in spite of all his arguments and a favorable committee report, the opposition from the cotton States proved too strong, and his application was denied. Whitney now had other interests. He was a great manufacturer of firearms, at New Haven, and as such we shall meet him again in a later chapter.

CHAPTER III
STEAM IN CAPTIVITY

For the beginnings of the enslavement of steam, that mighty giant whose work has changed the world we live in, we must return to the times of Benjamin Franklin. James Watt, the accredited father of the modern steam engine, was a contemporary of Franklin, and his engine was twenty-one years old when Franklin died. The discovery that steam could be harnessed and made to work is not, of course, credited to James Watt. The precise origin of that discovery is unknown. The ancient Greeks had steam engines of a sort, and steam engines of another sort were pumping water out of mines in England when James Watt was born. James Watt, however, invented and applied the first effective means by which steam came to serve mankind. And so the modern steam engine begins with him.

The story is old, of how this Scottish boy, James Watt, sat on the hearth in his mother's cottage, intently watching the steam rising from the mouth of the tea kettle, and of the great role which this boy afterwards assumed in the mechanical world. It was in 1763, when he was twenty-eight and had the appointment of mathematical-instrument maker to the University of Glasgow, that a model of Newcomen's steam pumping engine was brought into his shop for repairs. One can perhaps imagine the feelings with which James Watt, interested from his youth in mechanical and scientific instruments, particularly those which dealt with steam, regarded this Newcomen engine. Now his interest was vastly quickened. He set up the model and operated it, noticed how the alternate heating and cooling of its cylinder wasted power, and concluded, after some weeks of experiment, that, in order to make the engine practicable, the cylinder must be kept hot, "always as hot as the steam which entered it." Yet in order to condense the steam there must be a cooling of the vessel. The problem was to reconcile these two conditions.

At length the pregnant idea occurred to him—the idea of the separate condenser. It came to him on a Sunday afternoon in 1765, as he walked across Glasgow Green. If the steam were condensed in a vessel separate from the cylinder, it would be quite possible to keep the condensing vessel cool and the cylinder hot at the same time. Next morning Watt began to put his scheme to the test and found it practicable. He developed other ideas and applied them. So at last was born a steam engine that would work and multiply man's energies a thousandfold.

After one or two disastrous business experiences, such as fall to the lot of many great inventors, perhaps to test their perseverance, Watt associated himself with Matthew Boulton, a man of capital and of enterprise, owner of the Soho Engineering Works, near Birmingham. The firm of Boulton and Watt became famous, and James Watt lived till August 19, 1819—lived to see his steam engine the greatest single factor in the new industrial era that had dawned for English-speaking folk.

Boulton and Watt, however, though they were the pioneers, were by no means alone in the development of the steam engine. Soon there were rivals in the field with new types of engines. One of these was Richard Trevithick in England; another was Oliver Evans of Philadelphia. Both Trevithick and Evans invented the high-pressure engine. Evans appears to have applied the high pressure principle before Trevithick, and it has been said that Trevithick borrowed it from Evans, but Evans himself never said so, and it is more likely that each of these inventors worked it out independently. Watt introduced his steam to the cylinder at only slightly more than atmospheric pressure and clung tenaciously to the low-pressure theory all his life. Boulton and Watt, indeed, aroused by Trevithick's experiments in high-pressure engines, sought to have Parliament pass an act forbidding high pressure on the ground that the lives of the public were endangered. Watt lived long enough, however, to see the high-pressure steam engine come into general favor, not only in America but even in his own conservative country.

Less sudden, less dramatic, than that of the cotton gin, was the entrance of the steam engine on the American industrial stage, but not less momentous. The actions and reactions of steam in America provide the theme for an Iliad which some American Homer may one day write. They include the epic of the coal in the Pennsylvania hills, the epic of the ore, the epic of the railroad, the epic of the great city; and, in general, the subjugation of a continental wilderness to the service of a vast civilization.

The vital need of better transportation was uppermost in the thoughts of many Americans. It was seen that there could be no national unity in a country so far flung without means of easy intercourse between one group of Americans and another. The highroads of the new country were, for the most part, difficult even for the man on horseback, and worse for those who must travel by coach or post-chaise. Inland from the coast and away from the great rivers there were no roads of any sort; nothing but trails.

Highways were essential, not only for the permanent unity of the United States, but to make available the wonderful riches of the inland country, across the Appalachian barrier and around the Great Lakes, into which American pioneers had already made their way.

Those immemorial pathways, the great rivers, were the main avenues of traffic with the interior. So, of course, when men thought of improving transportation, they had in mind chiefly transportation by water; and that is why the earliest efforts of American inventors were applied to the means of improving traffic and travel by water and not by land.

The first men to spend their time in trying to apply steam power to the propulsion of a boat were contemporaries of Benjamin Franklin. Those who worked without Watt's engine could hardly succeed. One of the earliest of these was William Henry of Pennsylvania. Henry, in 1763, had the idea of applying power to paddle wheels, and constructed a boat, but his boat sank, and no result followed, unless it may be that John Fitch and Robert Fulton, both of whom were visitors at Henry's house, received some suggestions from him. James Rumsey of Maryland began experiments as early as 1774 and by 1786 had a boat that made four miles an hour against the current of the Potomac.

The most interesting of these early and unsuccessful inventors is John Fitch, who, was a Connecticut clockmaker living in Philadelphia. He was eccentric and irregular in his habits and quite ignorant of the steam engine. But he conceived the idea of a steamboat and set to work to make one. The record of Fitch's life is something of a tragedy. At the best he was an unhappy man and was always close to poverty. As a young man he had left his family because of unhappy domestic relations with his wife. One may find in the record of his undertakings which he left in the Philadelphia Library, to be opened thirty years after its receipt, these words: "I know of nothing so perplexing and vexatious to a man of feelings as a turbulent Wife and Steamboat building." But in spite of all his difficulties Fitch produced a steamboat, which plied regularly on the Delaware for several years and carried passengers. "We reigned Lord High Admirals of the Delaware; and no other boat in the River could hold its way with us," he wrote. "Thus has been effected by little Johnny Fitch and Harry Voight [one of his associates] one of the greatest and most useful arts that has ever been introduced into the world; and although the world and my country does not thank me for it, yet it gives me heartfelt satisfaction." The "Lord High Admirals of the Delaware," however, did not reign long. The steamboat needed improvement to make it pay; its backers lost patience and faith, and the inventor gave up the fight and retired into the fastnesses of the Kentucky wilderness, where he died.

The next inventor to struggle with the problem of the steamboat, with any approach to success, was John Stevens of Hoboken. His life was cast in a vastly different environment from that of John Fitch. He was a rich man, a man of family and of influence. His father's house—afterwards his own—-at 7 Broadway, facing Bowling Green—was one of the mansions of early New York, and his own summer residence on Castle Point, Hoboken, just across the Hudson, was one of the landmarks of the great river. For many years John Stevens crossed that river; most often in an open boat propelled by sail or by men at the oars. Being naturally of a mechanical turn, he sought to make the crossing easier. To his library were coming the prints that told of James Watt and the steam engine in England, and John Fitch's boat had interested him.

Robert Fulton's Clermont, of which we shall speak presently, was undoubtedly the pioneer of practicable steamboats. But the Phoenix, built by John Stevens, followed close on the Clermont. And its engines were built in America, while those of the Clermont had been imported from England. Moreover, in June, 1808, the Phoenix stood to sea, and made the first ocean voyage in the history of steam navigation. Because of a monopoly of the Hudson, which the New York Legislature had granted to Livingston and Fulton, Stevens was compelled to send his ship to the Delaware. Hence the trip out into the waters of the Atlantic, a journey that was not undertaken without trepidation. But, despite the fact that a great storm arose, the Phoenix made the trip in safety; and continued for many years thereafter to ply the Delaware between Philadelphia and Trenton.

Robert Fulton, like many and many another great inventor, from Leonardo da Vinci down to the present time, was also an artist. He was born November 14, 1765, at Little Britain, Lancaster County, Pennsylvania, of that stock which is so often miscalled "Scotch-Irish." He was only a child when his father died, leaving behind him a son who seems to have been much more interested in his own ideas than in his schoolbooks. Even in his childhood Robert showed his mechanical ability. There was a firm of noted gunsmiths in Lancaster, in whose shops he made himself at home and became expert in the use of tools. At the age of fourteen he applied his ingenuity to a heavy fishing boat and equipped it with paddle-wheels, which were turned by a crank, thus greatly lightening the labor of moving it.

At the age of seventeen young Fulton moved to Philadelphia and set up as a portrait painter. Some of the miniatures which he painted at this time are said to be very good. He worked hard, made many good friends, including Benjamin Franklin, and succeeded financially. He determined to go to Europe to study—if possible under his fellow Pennsylvanian,

Benjamin West, then rising into fame in London. The West and the Fulton families had been intimate, and Fulton hoped that West would take him as a pupil. First buying a farm for his mother with a part of his savings, he sailed for England in 1786, with forty guineas in his pocket. West received him not only as a pupil but as a guest in his house and introduced him to many of his friends. Again Fulton succeeded, and in 1791 two of his portraits were exhibited at the Royal Academy, and the Royal Society of British Artists hung four paintings by him.

Then came the commission which changed the course of Fulton's life. His work had attracted the notice of Viscount Courtenay, later Earl of Devon, and he was invited to Devonshire to paint that nobleman's portrait. Here he met Francis, third Duke of Bridgewater, the father of the English canal system, and his hardly less famous engineer, James Brindley, and also Earl Stanhope, a restless, inquiring spirit. Fulton the mechanic presently began to dominate Fulton the artist. He studied canals, invented a means of sawing marble in the quarries, improved the wheel for spinning flax, invented a machine for making rope, and a method of raising canal boats by inclined planes instead of locks. What money he made from these inventions we do not know, but somewhat later (1796) he speaks hopefully of an improvement in tanning. This same year he published a pamphlet entitled "A Treatise on the Improvement of Canal Navigation", copies of which were sent to Napoleon and President Washington.

Fulton went to France in 1797. To earn money he painted several portraits and a panorama of the Burning of Moscow. This panorama, covering the walls of a circular hall built especially for it, became very popular, and Fulton painted another. In Paris he formed a warm friendship with that singular American, Joel Barlow, soldier, poet, speculator, and diplomatist, and his wife, and for seven years lived in their house.

The long and complicated story of Fulton's sudden interest in torpedoes and submarine boats, his dealings with the Directory and Napoleon and with the British Admiralty does not belong here. His experiments and his negotiations with the two Governments occupied the greater part of his time for the years between 1797 and 1806. His expressed purpose was to make an engine of war so terrible that war would automatically be abolished. The world, however, was not ready for diving boats and torpedoes, nor yet for the end of war, and his efforts had no tangible results.*

** The submarine was the invention of David Bushnell, a Connecticut Yankee, whose "American Turtle" blew up at least one British vessel in the War of Independence and created much consternation among the King's ships in American waters.*

During all the years after 1793, at least, and perhaps earlier, the idea of the steamboat had seldom been out of his mind, but lack of funds and the greater urgency, as he thought, of the submarine prevented him from working seriously upon it. In 1801, however, Robert R. Livingston came to France as American Minister. Livingston had already made some unsuccessful experiments with the steamboat in the United States, and, in 1798, had received the monopoly of steam navigation on the waters of New York for twenty years, provided that he produced a vessel within twelve months able to steam four miles an hour. This grant had, of course, been forfeited, but might be renewed, Livingston thought.

Fulton and Livingston met, probably at Barlow's house, and, in 1802, drew up an agreement to construct a steamboat to ply between New York and Albany. Livingston agreed to advance five hundred dollars for experimentation in Europe. In this same year Fulton built a model and tested different means of propulsion, giving "the preference to a wheel on each side of the model."* The boat was built on the Seine, but proved too frail for the borrowed engine. A second boat was tried in August, 1803, and moved, though at a disappointingly slow rate of speed.

* *Fulton to Barlow, quoted in Sutcliffe, "Robert Fulton and*
the Clermont", p. 124.

Just at this time Fulton wrote ordering an engine from Boulton and Watt to be transported to America. The order was at first refused, as it was then the shortsighted policy of the British Government to maintain a monopoly of mechanical contrivances. Permission to export was given the next year, however, and the engine was shipped in 1805. It lay for some time in the New York Customs House. Meanwhile Fulton had studied the Watt engine on Symington's steamboat, the Charlotte Dundas, on the Forth and Clyde Canal, and Livingston had been granted a renewal of his monopoly of the waters of New York.

Fulton arrived at New York in 1806 and began the construction of the Clermont, so named after Livingston's estate on the Hudson. The building was done on the East River. The boat excited the jeers of passersby, who called it "Fulton's Folly." On Monday, August 17, 1807, the memorable first voyage was begun. Carrying a party of invited guests, the Clermont steamed off at one o'clock. Past the towns and villages along the Hudson, the boat moved steadily, black smoke rolling from her stack. Pine wood was the fuel. During the night, the sparks pouring from her funnel, the clanking of her machinery, and the splashing of the paddles frightened the animals in the woods and the occupants of the scattered houses along the banks. At one o'clock Tuesday the boat arrived at Clermont, 110 miles from New

York. After spending the night at Clermont, the voyage was resumed on Wednesday. Albany, forty miles away, was reached in eight hours, making a record of 150 miles in thirty-two hours. Returning to New York, the distance was covered in thirty hours. The steamboat was a success.

The boat was then laid up for two weeks while the cabins were boarded in, a roof built over the engine, and coverings placed over the paddle-wheels to catch the spray—all under Fulton's eye. Then the Clermont began regular trips to Albany, carrying sometimes a hundred passengers, making the round trip every four days, and continued until floating ice marked the end of navigation for the winter.

Why had Fulton succeeded where others had failed? There was nothing new in his boat. Every essential feature of the Clermont had been anticipated by one or other of the numerous experimenters before him. The answer seems to be that he was a better engineer than any of them. He had calculated proportions, and his hull and his engine were in relation. Then too, he had one of Watt's engines, undoubtedly the best at the time, and the unwavering support of Robert Livingston.

Fulton's restless mind was never still, but he did not turn capriciously from one idea to another. Though never satisfied, his new ideas were tested scientifically and the results carefully written down. Some of his notebooks read almost like geometrical demonstrations; and his drawings and plans were beautifully executed. Before his death in 1815 he had constructed or planned sixteen or seventeen boats, including boats for the Hudson, Potomac, and Mississippi rivers, for the Neva in Russia, and a steam vessel of war for the United States. He was a member of the commission on the Erie Canal, though he did not live to see that enterprise begun.

The mighty influence of the steamboat in the development of inland America is told elsewhere in this Series.* The steamboat has long since grown to greatness, but it is well to remember that the true ancestor of the magnificent leviathan of our own day is the Clermont of Robert Fulton.

Archer B. Hulbert, "The Paths of Inland Commerce".

The world today is on the eve of another great development in transportation, quite as revolutionary as any that have preceded. How soon will it take place? How long before Kipling's vision in "The Night Mail" becomes a full reality? How long before the air craft comes to play a great role in the world's transportation? We cannot tell. But, after looking at the nearest parallel in the facts of history, each of us may make his own guess. The airship appears now to be much farther advanced than the steamboat was for many years after Robert Fulton died. Already we have seen men ride the wind above the sea from the New World to the Old. Already United

States mails are regularly carried through the air from the Atlantic to the Golden Gate. It was twelve years after the birth of Fulton's Clermont, and four years after the inventor's death, before any vessel tried to cross the Atlantic under steam. This was in 1819, when the sailing packet Savannah, equipped with a ninety horsepower horizontal engine and paddle-wheels, crossed from Savannah to Liverpool in twenty-five days, during eighteen of which she used steam power. The following year, however, the engine was taken out of the craft. And it was not until 1833 that a real steamship crossed the Atlantic. This time it was the Royal William, which made a successful passage from Quebec to London. Four years more passed before the Great Western was launched at Bristol, the first steamship to be especially designed for transatlantic service, and the era of great steam liners began.

If steam could be made to drive a boat on the water, why not a wagon on the land?

History, seeking origins, often has difficulty when it attempts to discover the precise origin of an idea. "It frequently happens," said Oliver Evans, "that two persons, reasoning right on a mechanical subject, think alike and invent the same thing without any communication with each other."* It is certain, however, that one of the first, if not the first, protagonist of the locomotive in America was the same Oliver Evans, a truly great inventor for whom the world was not quite ready. The world has forgotten him. But he was the first engine builder in America, and one of the best of his day. He gave to his countrymen the high-pressure steam engine and new machinery for manufacturing flour that was not superseded for a hundred years.

Coleman Sellers, "Oliver Evans and His Inventions,"

"Journal of the Franklin Institute", July, 1886: vol. CXXII,

p. 16.

"Evans was apprenticed at the age of fourteen to a wheelwright. He was a thoughtful, studious boy, who devoured eagerly the few books to which he had access, even by the light of a fire of shavings, when denied a candle by his parsimonious master. He says that in 1779, when only seventeen years old, he began to contrive some method of propelling land carriages by other means than animal power; and that he thought of a variety of devices, such as using the force of the wind and treadles worked by men; but as they were evidently inadequate, was about to give up the problem as unsolvable for want of a suitable source of power, when he heard that some neighboring blacksmith's boys had stopped up the touch-hole of a gun barrel, put in some water, rammed down a tight wad, and, putting the breech into the smith's fire, the gun had discharged itself with a report like that of gunpowder. This immediately suggested to his fertile mind a new source of power, and he labored long to apply it, but without success, until there fell into his hands a book describing the old atmospheric steam engine of Newcomen, and he was at once struck with the fact that steam was only used to produce a vacuum while to him it seemed clear that the elastic power of the steam if applied directly to moving the piston, would be far more efficient. He soon satisfied himself that he could make steam wagons, but could convince no one else of this possibility."*

Coleman Sellers, "Oliver Evans and His Inventions,"

"Journal of the Franklin Institute", July, 1886: vol. CXXII,

p. 3.

Evans was then living in Delaware, where he was born, and where he later worked out his inventions in flour-milling machinery and invented and put into service the high-pressure steam engine. He appears to have moved to Philadelphia about 1790, the year of Franklin's death and of the Federal Patent Act; and, as we have seen, the third patent issued by the Government at Philadelphia was granted to him. About this time he became absorbed in the hard work of writing a book, the "Millwright and Miller's Guide", which he published in 1795, but at a heavy sacrifice to himself in time and money. A few years later he had an established engine works in Philadelphia and was making steam engines of his own type that performed their work satisfactorily.

The Oruktor Amphibolos, or Amphibious Digger, which came out of his shop in 1804, was a steamdriven machine made to the order of the Philadelphia Board of Health for dredging and cleaning the docks of the city. It was designed, as its name suggests, for service either in water or on shore. It propelled itself across the city to the river front, puffing and throwing off clouds of steam and making quite a sensation on the streets.

Evans had never forgotten his dream of the "steam wagon." His Oruktor had no sooner begun puffing than he offered to make for the Philadelphia and Lancaster Turnpike Company steamdriven carriages to take the place of their six-horse Conestoga wagons, promising to treble their profits. But the directors of the road were conservative men and his arguments fell on deaf ears.

In the same year Evans petitioned Congress for an extension of the patent on his flour-milling machinery, which was about to expire. He had derived little profit from this important invention, as the new machinery made its way very slowly, but every year more and more millers were using it and Evans received royalties from them. He felt sure that Congress would renew his patent, and, with great expectations for the future, he announced a new book in preparation by himself to be called "The Young Engineer's Guide". It was to give the most thorough treatment to the subject of the steam engine, with a profusion of drawings to illustrate the text. But Evans reckoned without the millers who were opposing his petition. Though they were profiting by his invention, they were unwilling to pay him anything, and they succeeded in having his bill in Congress defeated. It was a hard blow for the struggling author and inventor. His income cut off, he was obliged to reduce the scale of his book "and to omit many of the illustrations he had promised." He wrote the sad story into the name of the book. It came out under the title of "The Abortion of the Young Engineer's Guide".

Four years later, when Congress restored and extended his patent, Evans felt that better days were ahead, but, as said already, he was too far ahead of his time to be understood and appreciated. Incredulity, prejudice, and opposition were his portion as long as he lived. Nevertheless, he went on building good engines and had the satisfaction of seeing them in extensive use. His life came to an end as the result of what to him was the greatest possible tragedy. He was visiting New York City in 1819, when news came to him of the destruction by an incendiary of his beloved shops in Philadelphia. The shock was greater than he could bear. A stroke of apoplexy followed, from which he died.

The following prophecy, written by Oliver Evans and published in 1812, seventeen years before the practical use of the locomotive began, tells us something of the vision of this early American inventor:

"The time will come when people will travel in stages moved by steam engines from one city to another almost as fast as birds fly—fifteen to twenty miles an hour. Passing through the air with such velocity—changing the scenes in such rapid succession—will be the most exhilarating, delightful exercise. A carriage will set out from Washington in the morning, and the passengers will breakfast at Baltimore, dine in Philadelphia, and sup at New York the same day.

"To accomplish this, two sets of railways will be laid so nearly level as not in any place to deviate more than two degrees from a horizontal line, made of wood or iron, on smooth paths of broken stone or gravel, with a rail to guide the carriages so that they may pass each other in different directions and travel by night as well as by day; and the passengers will sleep in these stages as comfortably as they do now in steam stage-boats."*

Cited by Coleman Sellers, Ibid., p. 13.

Another early advocate of steam carriages and railways was John Stevens, the rich inventor of Hoboken, who figures in the story of the steamboat. In February, 1812, Stevens addressed to the commissioners appointed by the State of New York to explore a route for the Erie Canal an elaborate memoir calculated to prove that railways would be much more in the public interest than the proposed canal. He wrote at the same time to Robert R. Livingston (who, as well as Robert Fulton, his partner in the steamboat, was one of the commissioners) requesting his influence in favor of railways. Livingston, having committed himself to the steamboat and holding a monopoly of navigation on the waters of New York State, could hardly be expected to give a willing ear to a rival scheme, and no one then seems to have dreamed that both canal and railway would ultimately be needed. Livingston, however, was an enlightened statesman, one of the ablest men of his day. He had played a prominent part in the affairs of the Revolution and in the ratification of the Constitution; had known Franklin and Washington and had negotiated with Napoleon the Louisiana Purchase. His reply to Stevens is a good statement of the objections to the railway, as seen at the time, and of the public attitude towards it.

Robert R. Livingston to John Stevens

"Albany, 11th March, 1812.

"I did not, till yesterday, receive yours of the 5th of February; where it has loitered on the road I am at a loss to say. I had before read your very ingenious propositions as to the rail-way communication. I fear, however, on mature reflection, that they will be liable to serious objections, and ultimately more expensive than a canal. They must be double, so as to prevent the danger of two such heavy bodies meeting. The walls on which they are placed must at least be four feet below the surface, and three above, and must be clamped with iron, and even then, would hardly sustain so heavy a weight as you propose moving at the rate of four miles an hour on wheels. As to wood, it would not last a week; they must be covered with iron, and that too very thick and strong. The means of stopping these heavy carriages without a great shock, and of preventing them from running upon each other (for there would be many on the road at once) would be very difficult. In case of accidental stops, or the necessary stops to take wood and water &c many accidents would happen. The carriage of condensed water would be very troublesome. Upon the whole, I fear the expense would be much greater than that of canals, without being so convenient."*

John Stevens, "Documents Tending to Prove the Superior Advantages of Rail-Ways and Steam-Carriages over Canal Navigation" (1819). Reprinted in "The Magazine of History with Notes and Queries", Extra Number 54 (1917).

Stevens, of course, could not convince the commissioners. "The Communication from John Stevens, Esq.," was referred to a committee, who reported in March: "That they have considered the said communication with the attention due to a gentleman whose scientific researches and knowledge of mechanical powers entitle his opinions to great respect, and are sorry not to concur in them."

Stevens, however, kept up the fight. He published all the correspondence, hoping to get aid from Congress for his design, and spread his propaganda far and wide. But the War of 1812 soon absorbed the attention of the country. Then came the Erie Canal, completed in 1825, and the extension into the Northwest of the great Cumberland Road. From St. Louis steamboats churned their way up the Missouri, connecting with the Santa Fe Trail to the Southwest and the Oregon Trail to the far Northwest. Horses, mules, and oxen carried the overland travelers, and none yet dreamed of being carried on the land by steam.

Back East, however, and across the sea in England, there were a few dreamers. Railways of wooden rails, sometimes covered with iron, on which wagons were drawn by horses, were common in Great Britain; some were in use very early in America. And on these railways, or tramways, men were now experimenting with steam, trying to harness it to do the work of horses. In England, Trevithick, Blenkinsop, Ericsson, Stephenson, and others; in America, John Stevens, now an old man but persistent in his plans as ever and with able sons to help him, had erected a circular railway at Hoboken as early as 1826, on which he ran a locomotive at the rate of twelve miles an hour. Then in 1828 Horatio Allen, of the Delaware and Hudson Canal Company, went over to England and brought back with him the Stourbridge Lion. This locomotive, though it was not a success in practice, appears to have been the first to turn a wheel on a regular railway within the United States. It was a seven days' wonder in New York when it arrived in May, 1829. Then Allen shipped it to Honesdale, Pennsylvania, where the Delaware and Hudson Canal Company had a tramway to bring down coal from the mountains to the terminal of the canal. On the crude wooden rails of this tramway Allen placed the Stourbridge Lion and ran it successfully at the rate of ten miles an hour. But in actual service the Stourbridge Lion failed and was soon dismantled.

Pass now to Rainhill, England, and witness the birth of the modern locomotive, after all these years of labor. In the same year of 1829, on the morning of the 6th of October, a great crowd had assembled to see an extraordinary race—a race, in fact, without any parallel or precedent whatsoever. There were four entries but one dropped out, leaving three: The Novelty, John Braithwaite and John Ericsson; The Sanspareil, Timothy Hackworth; The Rocket, George and Robert Stephenson. These were not horses; they were locomotives. The directors of the London and Manchester

Railway had offered a prize of five hundred pounds for the best locomotive, and here they were to try the issue.

The contest resulted in the triumph of Stephenson's Rocket. The others fell early out of the race. The Rocket alone met all the requirements and won the prize. So it happened that George Stephenson came into fame and has ever since lived in popular memory as the father of the locomotive. There was nothing new in his Rocket, except his own workmanship. Like Robert Fulton, he appears to have succeeded where others failed because he was a sounder engineer, or a better combiner of sound principles into a working, whole, than any of his rivals.

Across the Atlantic came the news of Stephenson's remarkable success. And by this time railroads were beginning in various parts of the United States: the Mohawk and Hudson, from Albany to Schenectady; the Baltimore and Ohio; the Charleston and Hamburg in South Carolina; the Camden and Amboy, across New Jersey. Horses, mules, and even sails, furnished the power for these early railroads. It can be imagined with what interest the owners of these roads heard that at last a practicable locomotive was running in England.

This news stimulated the directors of the Baltimore and Ohio to try the locomotive. They had not far to go for an experiment, for Peter Cooper, proprietor of the Canton Iron Works in Baltimore, had already designed a small locomotive, the Tom Thumb. This was placed on trial in August, 1830, and is supposed to have been the first American-built locomotive to do work on rails, though nearly coincident with it was the Best Friend of Charleston, built by the West Point Foundry, New York, for the Charleston and Hamburg Railroad. It is often difficult, as we have seen, to say which of two or several things was first. It appears as though the little Tom Thumb was the first engine built in America, which actually pulled weight on a regular railway, while the much larger Best Friend was the first to haul cars in regular daily service.

The West Point Foundry followed its first success with the West Point, which also went into service on the Charleston and Hamburg Railroad, and then built for the newly finished Mohawk and Hudson (the first link in the New York Central Lines) the historic De Witt Clinton. This primitive locomotive and the cars it drew may be seen today in the Grand Central Station in New York.

Meanwhile, the Stevens brothers, sons of John Stevens, were engaged in the construction of the Camden and Amboy Railroad. The first locomotive to operate on this road was built in England by George Stephenson. This was the John Bull, which arrived in the summer of 1831 and at once went to work. The John Bull was a complete success and had a distinguished career. Sixty-two years old, in 1893, it went to Chicago, to the Columbian Exposition, under its own steam. The John Bull occupies a place today in the National Museum at Washington.

With the locomotive definitely accepted, men began to turn their minds towards its improvement and development, and locomotive building soon became a leading industry in America. At first the British types and patterns were followed, but it was not long before American designers began to depart from the British models and to evolve a distinctively American type. In the development of this type great names have been written into the industrial history of America, among which the name of Matthias Baldwin of Philadelphia probably ranks first. But there have been hundreds of great workers in this field. From Stephenson's Rocket and the little Tom Thumb of Peter Cooper, to the powerful "Mallets" of today, is a long distance—not spanned in ninety years save by the genius and restless toil of countless brains and hands.

If the locomotive could not remain as it was left by Stephenson and Cooper, neither could the stationary steam engine remain as it was left by James Watt and Oliver Evans. Demands increasing and again increasing, year after year, forced the steam engine to grow in order to meet its responsibilities. There were men living in Philadelphia in 1876, who had known Oliver Evans personally; at least one old man at the Centennial Exhibition had himself seen the Oruktor Amphibolos and recalled the consternation it had caused on the streets of the city in 1804. It seemed a far cry back to the Oruktor from the great and beautiful engine, designed by George Henry Corliss, which was then moving all the vast machinery of the Centennial Exhibition. But since then achievements in steam have dwarfed even the great work of Corliss. And to do a kind of herculean task that was hardly dreamed of in 1876 another type of engine has made its entrance: the steam turbine, which sends its awful energy, transformed into electric current, to light a million lamps or to turn ten thousand wheels on distant streets and highways.

CHAPTER IV
SPINDLE, LOOM, AND NEEDLE
IN NEW ENGLAND

The major steps in the manufacture of clothes are four: first to harvest and clean the fiber or wool; second, to card it and spin it into threads; third, to weave the threads into cloth; and, finally to fashion and sew the cloth into clothes. We have already seen the influence of Eli Whitney's cotton gin on the first process, and the series of inventions for spinning and weaving, which so profoundly changed the textile industry in Great Britain, has been mentioned. It will be the business of this chapter to tell how spinning and weaving machinery was introduced into the United States and how a Yankee inventor laid the keystone of the arch of clothing machinery by his invention of the sewing machine.

Great Britain was determined to keep to herself the industrial secrets she had gained. According to the economic beliefs of the eighteenth century, which gave place but slowly to the doctrines of Adam Smith, monopoly rather than cheap production was the road to success. The laws therefore forbade the export of English machinery or drawings and specifications by which machines might be constructed in other countries. Some men saw a vast prosperity for Great Britain, if only the mystery might be preserved.

Meanwhile the stories of what these machines could do excited envy in other countries, where men desired to share in the industrial gains. And, even before Eli Whitney's cotton gin came to provide an abundant supply of raw material, some Americans were struggling to improve the old hand loom, found in every house, and to make some sort of a spinning machine to replace the spinning wheel by which one thread at a time was laboriously spun.

East Bridgewater, Massachusetts, was the scene of one of the earliest of these experiments. There in 1786 two Scotchmen, who claimed to understand Arkwright's mechanism, were employed to make spinning machines, and about the same time another attempt was made at Beverly. In both instances the experiments were encouraged by the State and assisted with grants of money. The machines, operated by horse power, were crude, and the product was irregular and unsatisfactory. Then three men at Providence, Rhode Island, using drawings of the Beverly machinery, made machines having thirty-two spindles which worked indifferently. The attempt to run them by water power failed, and they were sold to Moses Brown of Pawtucket, who with his partner, William Almy, had mustered an army of hand-loom weavers in 1790, large enough to produce nearly eight thousand yards of cloth in that year. Brown's need of spinning machinery, to provide his weavers with yarn, was very great; but these machines he had bought would not run, and in 1790 there was not a single successful power-spinner in the United States.

Meanwhile Benjamin Franklin had come home, and the Pennsylvania Society for the Encouragement of Manufactures and Useful Arts was offering prizes for inventions to improve the textile industry. And in Milford, England, was a young man named Samuel Slater, who, on hearing that inventive genius was munificently rewarded in America, decided to migrate to that country. Slater at the age of fourteen had been apprenticed to Jedediah Strutt, a partner of Arkwright. He had served both in the counting-house and the mill and had had every opportunity to learn the whole business.

Soon after attaining his majority, he landed in New York, November, 1789, and found employment. From New York he wrote to Moses Brown of Pawtucket, offering his services, and that old Quaker, though not giving him much encouragement, invited him to Pawtucket to see whether he could run the spindles which Brown had bought from the men of Providence. "If thou canst do what thou sayest," wrote Brown, "I invite thee to come to Rhode Island."

Arriving in Pawtucket in January, 1790, Slater pronounced the machines worthless, but convinced Almy and Brown that he knew his business, and they took him into partnership. He had no drawings or models of the English machinery, except such as were in his head, but he proceeded to build machines, doing much of the work himself. On December 20, 1790, he had ready carding, drawing, and roving machines and seventy-two spindles in two frames. The water-wheel of an old fulling mill furnished the power—and the machinery ran.

Here then was the birth of the spinning industry in the United States. The "Old Factory," as it was to be called for nearly a hundred years, was built at Pawtucket in 1793. Five years later Slater and others built a second mill, and in 1806, after Slater had brought out his brother to share his prosperity, he built another. Workmen came to work for him solely to learn his machines, and then left him to set up for themselves. The knowledge he had brought soon became widespread. Mills were built not only in New England but in other States. In 1809 there were sixty-two spinning mills in operation in the country, with thirty-one thousand spindles; twenty-five more mills were building or projected, and the industry was firmly established in the United States. The yarn was sold to housewives for domestic use or else to professional weavers who made cloth for sale. This practice was continued for years, not only in New England, but also in those other parts of the country where spinning machinery had been introduced.

By 1810, however, commerce and the fisheries had produced considerable fluid capital in New England which was seeking profitable employment, especially as the Napoleonic Wars interfered with American shipping; and since Whitney's gins in the South were now piling up mountains of raw cotton, and Slater's machines in New England were making this cotton into yarn, it was inevitable that the next step should be the power loom, to convert the yarn into cloth. So Francis Cabot Lowell, scion of the New England family of that name, an importing merchant of Boston, conceived the idea of establishing weaving mills in Massachusetts. On a visit to Great Britain in 1811, Lowell met at Edinburgh Nathan Appleton, a fellow merchant of Boston, to whom he disclosed his plans and announced his intention of going to Manchester to gain all possible information concerning the new industry. Two years afterwards, according to Appleton's account, Lowell and his brother-in-law, Patrick T. Jackson, conferred with Appleton at the Stock Exchange in Boston. They had decided, they said, to set up a cotton factory at Waltham and invited Appleton to join them in the adventure, to which he readily consented. Lowell had not been able to obtain either drawings or model in Great Britain, but he had nevertheless designed a loom and had completed a model which seemed to work.

The partners took in with them Paul Moody of Amesbury, an expert machinist, and by the autumn of 1814 looms were built and set up at Waltham. Carding, drawing, and roving machines were also built and installed in the mill, these machines gaining greatly, at Moody's expert hands, over their American rivals. This was the first mill in the United States, and one of the first in the world, to combine under one roof all the operations necessary to convert raw fiber into cloth, and it proved a success. Lowell, says his partner Appleton, "is entitled to the credit for having introduced the new system in the cotton manufacture." Jackson and Moody "were men of unsurpassed talent," but Lowell "was the informing soul, which gave direction and form to the whole proceeding."

The new enterprise was needed, for the War of 1812 had cut off imports. The beginnings of the protective principle in the United States tariff are now to be observed. When the peace came and Great Britain began to dump goods in the United States, Congress, in 1816, laid a minimum duty of six and a quarter cents a yard on imported cottons; the rate was raised in 1824 and again in 1828. It is said that Lowell was influential in winning the support of John C. Calhoun for the impost of 1816.

Lowell died in 1817, at the early age of forty-two, but his work did not die with him. The mills he had founded at Waltham grew exceedingly prosperous under the management of Jackson; and it was not long before Jackson and his partners Appleton and Moody were seeking wider opportunities. By 1820 they were looking for a suitable site on which to build new mills, and their attention was directed to the Pawtucket Falls, on the Merrimac River. The land about this great water power was owned by the Pawtucket Canal Company, whose canal, built to improve the navigation of the Merrimac, was not paying satisfactory profits. The partners proceeded to acquire the stock of this company and with it the land necessary for their purpose, and in December, 1821, they executed Articles of Association for the Merrimac Manufacturing Company, admitting some additional partners, among them Kirk Boott who was to act as resident agent and manager of the new enterprise, since Jackson could not leave his duties at Waltham.

The story of the enterprise thus begun forms one of the brightest pages in the industrial history of America; for these partners had the wisdom and foresight to make provision at the outset for the comfort and well-being of their operatives. Their mill hands were to be chiefly girls drawn from the rural population of New England, strong and intelligent young women, of whom there were at that time great numbers seeking employment, since household manufactures had come to be largely superseded by factory goods. And one of the first questions which the partners considered was whether the change from farm to factory life would effect for the worse the character of these girls. This, says Appleton, "was a matter of deep interest. The operatives in the manufacturing cities of Europe were notoriously of the lowest character for intelligence and morals. The question therefore arose, and was deeply considered, whether this degradation was the result of the peculiar occupation or of other and distinct causes. We could not perceive why this peculiar description of labor should vary in its effects upon character from all other occupations." And so we find the partners voting money, not only for factory buildings and machinery, but for comfortable boardinghouses for the girls, and planning that these boardinghouses should have "the most efficient guards," that they should be in "charge of respectable women, with every provision for religious worship." They voted nine thousand dollars for a church building and further sums later for a library and a hospital.

The wheels of the first mill were started in September, 1823. Next year the partners petitioned the Legislature to have their part of the township set off to form a new town. One year later still they erected three new mills; and in another year (1826) the town of Lowell was incorporated.

The year 1829 found the Lowell mills in straits for lack of capital, from which, however, they were promptly relieved by two great merchants of Boston, Amos and Abbott Lawrence, who now became partners in the business and who afterwards founded the city named for them farther down on the Merrimac River.

The story of the Lowell cotton factories, for twenty years, more or less, until the American girls operating the machines came to be supplanted by French Canadians and Irish, is appropriately summed up in the title of a book which describes the factory life in Lowell during those years. The title of this book is "An Idyl of Work" and it was written by Lucy Larcom, who was herself one of the operatives and whose mother kept one of the corporation boarding-houses. And Lucy Larcom was not the only one of the Lowell "factory girls" who took to writing and lecturing. There were many others, notably, Harriet Hanson (later Mrs. W. S. Robinson), Harriot Curtis ("Mina Myrtle"), and Harriet Farley; and many of the "factory girls" married men who became prominent in the world. There was no thought among them that there was anything degrading in factory work. Most of the girls came from the surrounding farms, to earn money for a trousseau, to send a brother through college, to raise a mortgage, or to enjoy the society of their fellow workers, and have a good time in a quiet, serious way, discussing the sermons and lectures they heard and the books they read in their leisure hours. They had numerous "improvement circles" at which contributions of the members in both prose and verse were read and discussed. And for several years they printed a magazine, "The Lowell Offering", which was entirely written and edited by girls in the mills.

Charles Dickens visited Lowell in the winter of 1842 and recorded his impressions of what he saw there in the fourth chapter of his "American Notes". He says that he went over several of the factories, "examined them in every part; and saw them in their ordinary working aspect, with no preparation of any kind, or departure from their ordinary every-day proceedings"; that the girls "were all well dressed: and that phrase necessarily includes extreme cleanliness. They had serviceable bonnets, good warm cloaks, and shawls.... Moreover, there were places in the mill in which they could deposit these things without injury; and there were conveniences for washing. They were healthy in appearance, many of them remarkably so, and had the manners and deportment of young women; not of degraded brutes of burden." Dickens continues: "The rooms in which they worked were as well ordered as themselves. In the windows of some there were green plants, which were trained to shade the glass; in all, there was as much fresh air, cleanliness, and comfort as the nature of the occupation would possibly admit of." Again: "They reside in various boarding-houses near at hand. The owners of the mills are particularly careful to allow no persons to enter upon the possession of these houses, whose characters have not undergone the most searching and thorough enquiry." Finally, the author announces that he will state three facts which he thinks will startle his English readers: "Firstly, there is a joint-stock piano in a great many of the boarding-houses. Secondly, nearly all these young ladies subscribe to circulating libraries. Thirdly, they have got up among themselves a periodical called 'The Lowell Offering'... whereof I brought away from Lowell four hundred good solid pages, which I have read from beginning to end." And: "Of the merits of the 'Lowell Offering' as a literary production, I will only observe, putting entirely out of sight the fact of the articles having been written by these girls after the arduous labors of the day, that it will compare advantageously with a great many English Annuals."

The efficiency of the New England mills was extraordinary. James Montgomery, an English cotton manufacturer, visited the Lowell mills two years before Dickens and wrote after his inspection of them that they produced "a greater quantity of yarn and cloth from each spindle and loom (in a given time) than was produced by any other factories, without exception in the world." Long before that time, of course, the basic type of loom had changed from that originally introduced, and many New England inventors had been busy devising improved machinery of all kinds.

Such were the beginnings of the great textile mills of New England. The scene today is vastly changed. Productivity has been multiplied by invention after invention, by the erection of mill after mill, and by the employment of thousands of hands in place of hundreds. Lowell as a textile center has long been surpassed by other cities. The scene in Lowell itself is vastly changed. If Charles Dickens could visit Lowell today, he would hardly recognize in that city of modern factories, of more than a hundred thousand people, nearly half of them foreigners, the Utopia of 1842 which he saw and described.

The cotton plantations in the South were flourishing, and Whitney's gins were cleaning more and more cotton; the sheep of a thousand hills were giving wool; Arkwright's machines in England, introduced by Slater into New England, were spinning the cotton and wool into yarn; Cartwright's looms in England and Lowell's improvements in New England were weaving the yarn into cloth; but as yet no practical machine had been invented to sew the cloth into clothes.

There were in the United States numerous small workshops where a few tailors or seamstresses, gathered under one roof, laboriously sewed garments together, but the great bulk of the work, until the invention of the sewing machine, was done by the wives and daughters of farmers and sailors in the villages around Boston, New York, and Philadelphia. In these cities the garments were cut and sent out to the dwellings of the poor to be sewn. The wages of the laborers were notoriously inadequate, though probably better than in England. Thomas Hood's ballad The Song of the Shirt, published in 1843, depicts the hardships of the English woman who strove to keep body and soul together by means of the needle:

With fingers weary and worn,
With eyelids heavy and red,
A woman sat in unwomanly rags,
Plying her needle and thread.

Meanwhile, as Hood wrote and as the whole English people learned by heart his vivid lines, as great ladies wept over them and street singers sang them in the darkest slums of London, a man, hungry and ill-clad, in an attic in faraway Cambridge, Massachusetts, was struggling to put into metal an idea to lighten the toil of those who lived by the needle. His name was Elias Howe and he hailed from Eli Whitney's old home, Worcester County, Massachusetts. There Howe was born in 1819. His father was an unsuccessful farmer, who also had some small mills, but seems to have succeeded in nothing he undertook.

Young Howe led the ordinary life of a New England country boy, going to school in winter and working about the farm until the age of sixteen, handling tools every day, like any farmer's boy of the time. Hearing of high wages and interesting work in Lowell, that growing town on the Merrimac, he went there in 1835 and found employment; but two years later, when the panic of 1837 came on, he left Lowell and went to work in a machine shop in Cambridge. It is said that, for a time, he occupied a room with his cousin, Nathaniel P. Banks, who rose from bobbin boy in a cotton mill to Speaker of the United States House of Representatives and Major-General in the Civil War.

Next we hear of Howe in Boston, working in the shop of Ari Davis, an eccentric maker and repairer of fine machinery. Here the young mechanic heard of the desirability of a sewing machine and began to puzzle over the problem. Many an inventor before him had attempted to make sewing machines and some had just fallen short of success. Thomas Saint, an Englishman, had patented one fifty years earlier; and about this very time a Frenchman named Thimmonier was working eighty sewing machines making army uniforms, when needle workers of Paris, fearing that the bread was to be taken from them, broke into his workroom and destroyed the machines. Thimmonier tried again, but his machine never came into general use. Several patents had been issued on sewing machines in the United States, but without any practical result. An inventor named Walter Hunt had discovered the principle of the lock-stitch and had built a machine but had wearied of his work and abandoned his invention, just as success was in sight. But Howe knew nothing of any of these inventors. There is no evidence that he had ever seen the work of another.

The idea obsessed him to such an extent that he could do no other work, and yet he must live. By this time he was married and had children, and his wages were only nine dollars a week. Just then an old schoolmate, George Fisher, agreed to support his family and furnish him with five hundred dollars for materials and tools. The attic in Fisher's house in Cambridge was Howe's workroom. His first efforts were failures, but all at once the idea of the lock-stitch came to him. Previously all machines (except Hunt's, which was unknown, not having even been patented) had used the chainstitch, wasteful of thread and easily unraveled. The two threads of the lockstitch cross in the materials joined together, and the lines of stitches show the same on both sides. In short, the chainstitch is a crochet or knitting stitch, while the lockstitch is a weaving stitch. Howe had been working at night and was on his way home, gloomy and despondent, when this idea dawned on his mind, probably rising out of his experience in the cotton mill. The shuttle would be driven back and forth as in a loom, as he had seen it thousands of times, and passed through a loop of thread which the curved needle would throw out on the other side of the cloth; and the cloth would be fastened to the machine vertically by pins. A curved arm would ply the needle with the motion of a pick-axe. A handle attached to the fly-wheel would furnish the power.

On that design Howe made a machine which, crude as it was, sewed more rapidly than five of the swiftest needle workers. But apparently to no purpose. His machine was too expensive, it could sew only a straight seam, and it might easily get out of order. The needle workers were opposed, as they have generally been, to any sort of laborsaving machinery, and there was no manufacturer willing to buy even one machine at the price Howe asked, three hundred dollars.

Howe's second model was an improvement on the first. It was more compact and it ran more smoothly. He had no money even to pay the fees necessary to get it patented. Again Fisher came to the rescue and took Howe and his machine to Washington, paying all the expenses, and the patent was issued in September, 1846. But, as the machine still failed to find buyers,

Fisher gave up hope. He had invested about two thousand dollars which seemed gone forever, and he could not, or would not, invest more. Howe returned temporarily to his father's farm, hoping for better times.

Meanwhile Howe had sent one of his brothers to London with a machine to see if a foothold could be found there, and in due time an encouraging report came to the destitute inventor. A corsetmaker named Thomas had paid two hundred and fifty pounds for the English rights and had promised to pay a royalty of three pounds on each machine sold. Moreover, Thomas invited the inventor to London to construct a machine especially for making corsets. Howe went to London and later sent for his family. But after working eight months on small wages, he was as badly off as ever, for, though he had produced the desired machine, he quarrelled with Thomas and their relations came to an end.

An acquaintance, Charles Inglis, advanced Howe a little money while he worked on another model. This enabled Howe to send his family home to America, and then, by selling his last model and pawning his patent rights, he raised enough money to take passage himself in the steerage in 1848, accompanied by Inglis, who came to try his fortune in the United States.

Howe landed in New York with a few cents in his pocket and immediately found work. But his wife was dying from the hardships she had suffered, due to stark poverty. At her funeral, Howe wore borrowed clothes, for his only suit was the one he wore in the shop.

Then, soon after his wife had died, Howe's invention came into its own. It transpired presently that sewing machines were being made and sold and that these machines were using the principles covered by Howe's patent. Howe found an ally in George W. Bliss, a man of means, who had faith in the machine and who bought out Fisher's interest and proceeded to prosecute infringers. Meanwhile Howe went on making machines—he produced fourteen in New York during 1850—and never lost an opportunity to show the merits of the invention which was being advertised and brought to notice by the activities of some of the infringers, particularly by Isaac M. Singer, the best business man of them all. Singer had joined hands with Walter Hunt and Hunt had tried to patent the machine which he had abandoned nearly twenty years before.

The suits dragged on until 1854, when the case was decisively settled in Howe's favor. His patent was declared basic, and all the makers of sewing machines must pay him a royalty of twenty-five dollars on every machine. So Howe woke one morning to find himself enjoying a large income, which in time rose as high as four thousand dollars a week, and he died in 1867 a rich man.

Though the basic nature of Howe's patent was recognized, his machine was only a rough beginning. Improvements followed, one after another, until the sewing machine bore little resemblance to Howe's original. John Bachelder introduced the horizontal table upon which to lay the work. Through an opening in the table, tiny spikes in an endless belt projected and pushed the work for ward continuously. Allan B. Wilson devised a rotary hook carrying a bobbin to do the work of the shuttle, and also the small serrated bar which pops up through the table near the needle, moves forward a tiny space, carrying the cloth with it, drops down just below the upper surface of the table, and returns to its starting point, to repeat over and over again this series of motions. This simple device brought its owner a fortune. Isaac M. Singer, destined to be the dominant figure of the industry, patented in 1851 a machine stronger than any of the others and with several valuable features, notably the vertical presser foot held down by a spring; and Singer was the first to adopt the treadle, leaving both hands of the operator free to manage the work. His machine was good, but, rather than its surpassing merits, it was his wonderful business ability that made the name of Singer a household word.

By 1856 there were several manufacturers in the field, threatening war on each other. All men were paying tribute to Howe, for his patent was basic, and all could join in fighting him, but there were several other devices almost equally fundamental, and even if Howe's patents had been declared void it is probable that his competitors would have fought quite as fiercely among themselves. At the suggestion of George Gifford, a New York attorney, the leading inventors and manufacturers agreed to pool their inventions and to establish a fixed license fee for the use of each. This "combination" was composed of Elias Howe, Wheeler and Wilson, Grover and Baker, and I. M. Singer, and dominated the field until after 1877, when the majority of the basic patents expired. The members manufactured sewing machines and sold them in America and Europe. Singer introduced the installment plan of sale, to bring the machine within reach of the poor, and the sewing machine agent, with a machine or two on his wagon, drove through every small town and country district, demonstrating and selling. Meanwhile the price of the machines steadily fell, until it seemed that Singer's slogan, "A machine in every home!" was in a fair way to be realized, had not another development of the sewing machine intervened.

This was the development of the ready-made clothing industry. In the earlier days of the nation, though nearly all the clothing was of domestic manufacture, there were tailors and seamstresses in all the towns and many of the villages, who made clothing to order. Sailors coming ashore sometimes needed clothes at once, and apparently a merchant of New Bedford was the first to keep a stock on hand. About 1831, George Opdyke, later Mayor of New York, began the manufacture of clothing on Hudson Street, which he sold largely through a store in New Orleans. Other firms began to reach out for this Southern trade, and it became important. Southern planters bought clothes not only for their slaves but for their families. The development of California furnished another large market. A shirt factory was established, in 1832, on Cherry and Market Streets, New York. But not until the coming of the power-driven sewing machine could there be any factory production of clothes on a large scale. Since then the clothing industry has become one of the most important in the country. The factories have steadily improved their models and materials, and at the present day only a negligible fraction of the people of the United States wear clothes made to their order.

The sewing machine today does many things besides sewing a seam. There are attachments which make buttonholes, darn, embroider, make ruffles or hems, and dozens of other things. There are special machines for every trade, some of which deal successfully with refractory materials.

The Singer machine of 1851 was strong enough to sew leather and was almost at once adopted by the shoemakers. These craftsmen flourished chiefly in Massachusetts, and they had traditions reaching back at least to Philip Kertland, who came to Lynn in 1636 and taught many apprentices. Even in the early days before machinery, division of labor was the rule in the shops of Massachusetts. One workman cut the leather, often tanned on the premises; another sewed the uppers together, while another sewed on the soles. Wooden pegs were invented in 1811 and came into common use about 1815 for the cheaper grades of shoes: Soon the practice of sending out the uppers to be done by women in their own homes became common. These women were wretchedly paid, and when the sewing machine came to do the work better than it could be done by hand, the practice of "putting out" work gradually declined.

That variation of the sewing machine which was to do the more difficult work of sewing the sole to the upper was the invention of a mere boy, Lyman R. Blake. The first model, completed in 1858, was imperfect, but Blake was able to interest Gordon McKay, of Boston, and three years of patient experimentation and large expenditure followed. The McKay sole-sewing machine, which they produced, came into use, and for twenty-one years was used almost universally both in the United States and Great

Britain. But this, like all the other useful inventions, was in time enlarged and greatly improved, and hundreds of other inventions have been made in the shoe industry. There are machines to split leather, to make the thickness absolutely uniform, to sew the uppers, to insert eyelets, to cut out heel tops, and many more. In fact, division of labor has been carried farther in the making of shoes than in most industries, for there are said to be about three hundred separate operations in making a pair of shoes.

From small beginnings great industries have grown. It is a far cry from the slow, clumsy machine of Elias Howe, less than three-quarters of a century ago, to the great factories of today, filled with special models, run at terrific speed by electric current, and performing tasks which would seem to require more than human intelligence and skill.

CHAPTER V
THE AGRICULTURAL REVOLUTION

The Census of 1920 shows that hardly thirty per cent of the people are today engaged in agriculture, the basic industry of the United States, as compared with perhaps ninety per cent when the nation began. Yet American farmers, though constantly diminishing in proportion to the whole population, have always been, and still are, able to feed themselves and all their fellow Americans and a large part of the outside world as well. They bring forth also not merely foodstuffs, but vast quantities of raw material for manufacture, such as cotton, wool, and hides. This immense productivity is due to the use of farm machinery on a scale seen nowhere else in the world. There is still, and always will be, a good deal of hard labor on the farm. But invention has reduced the labor and has made possible the carrying on of this vast industry by a relatively small number of hands.

The farmers of Washington's day had no better tools than had the farmers of Julius Caesar's day; in fact, the Roman ploughs were probably superior to those in general use in America eighteen centuries later. "The machinery of production," says Henry Adams, "showed no radical difference from that familiar in ages long past. The Saxon farmer of the eighth century enjoyed most of the comforts known to Saxon farmers of the eighteenth."* One type of plough in the United States was little more than a crooked stick with an iron point attached, sometimes with rawhide, which simply scratched the ground. Ploughs of this sort were in use in Illinois as late as 1812. There were a few ploughs designed to turn a furrow, often simply heavy chunks of tough wood, rudely hewn into shape, with a wrought-iron point clumsily attached. The moldboard was rough and the curves of no two were alike. Country blacksmiths made ploughs only on order and few had patterns. Such ploughs could turn a furrow in soft ground if the oxen were strong enough—but the friction was so great that three men and four or six oxen were required to turn a furrow where the sod was tough.

* *"History of the United States", vol. I, p. 16.*

Thomas Jefferson had worked out very elaborately the proper curves of the moldboard, and several models had been constructed for him. He was, however, interested in too many things ever to follow any one to the end, and his work seems to have had little publicity. The first real inventor of a practicable plough was Charles Newbold, of Burlington County, New Jersey, to whom a patent for a cast-iron plough was issued in June, 1797. But the farmers would have none of it. They said it "poisoned the soil" and fostered the growth of weeds. One David Peacock received a patent in 1807, and two others later. Newbold sued Peacock for infringement and recovered damages. Pieces of Newbold's original plough are in the museum of the New York Agricultural Society at Albany.

Another inventor of ploughs was Jethro Wood, a blacksmith of Scipio, New York, who received two patents, one in 1814 and the other in 1819. His plough was of cast iron, but in three parts, so that a broken part might be renewed without purchasing an entire plough. This principle of standardization marked a great advance. The farmers by this time were forgetting their former prejudices, and many ploughs were sold. Though Wood's original patent was extended, infringements were frequent, and he is said to have spent his entire property in prosecuting them.

In clay soils these ploughs did not work well, as the more tenacious soil stuck to the iron moldboard instead of curling gracefully away. In 1833, John Lane, a Chicago blacksmith, faced a wooden moldboard with an old steel saw. It worked like magic, and other blacksmiths followed suit to such an extent that the demand for old saws became brisk. Then came John Deere, a native of Vermont, who settled first in Grand Detour, and then in Moline, Illinois. Deere made wooden ploughs faced with steel, like other blacksmiths, but was not satisfied with them and studied and experimented to find the best curves and angles for a plough to be used in the soils around him. His ploughs were much in demand, and his need for steel led him to have larger and larger quantities produced for him, and the establishment which still bears his name grew to large proportions.

Another skilled blacksmith, William Parlin, at Canton, Illinois, began making ploughs about 1842, which he loaded upon a wagon and peddled through the country. Later his establishment grew large. Another John Lane, a son of the first, patented in 1868 a "soft-center" steel plough. The hard but brittle surface was backed by softer and more tenacious metal, to reduce the breakage. The same year James Oliver, a Scotch immigrant who had settled at South Bend, Indiana, received a patent for the "chilled plough." By an ingenious method the wearing surfaces of the casting were cooled more quickly than the back. The surfaces which came in contact with the soil had a hard, glassy surface, while the body of the plough was of tough iron. From small beginnings Oliver's establishment grew great, and the Oliver Chilled

Plow Works at South Bend is today one of the largest and most favorably known privately owned industries in the United States.

From the single plough it was only a step to two or more ploughs fastened together, doing more work with approximately the same man power. The sulky plough, on which the ploughman rode, made his work easier, and gave him great control. Such ploughs were certainly in use as early as 1844, perhaps earlier. The next step forward was to substitute for horses a traction engine. Today one may see on thousands of farms a tractor pulling six, eight, ten, or more ploughs, doing the work better than it could be done by an individual ploughman. On the "Bonanza" farms of the West a fifty horsepower engine draws sixteen ploughs, followed by harrows and a grain drill, and performs the three operations of ploughing, harrowing, and planting at the same time and covers fifty acres or more in a day.

The basic ideas in drills for small grains were successfully developed in Great Britain, and many British drills were sold in the United States before one was manufactured here. American manufacture of these drills began about 1840. Planters for corn came somewhat later. Machines to plant wheat successfully were unsuited to corn, which must be planted less profusely than wheat.

The American pioneers had only a sickle or a scythe with which to cut their grain. The addition to the scythe of wooden fingers, against which the grain might lie until the end of the swing, was a natural step, and seems to have been taken quite independently in several places, perhaps as early as 1803. Grain cradles are still used in hilly regions and in those parts of the country where little grain is grown.

The first attempts to build a machine to cut grain were made in England and Scotland, several of them in the eighteenth century; and in 1822 Henry Ogle, a schoolmaster in Rennington, made a mechanical reaper, but the opposition of the laborers of the vicinity, who feared loss of employment, prevented further development. In 1826, Patrick Bell, a young Scotch student, afterward a Presbyterian minister, who had been moved by the fatigue of the harvesters upon his father's farm in Argyllshire, made an attempt to lighten their labor. His reaper was pushed by horses; a reel brought the grain against blades which opened and closed like scissors, and a traveling canvas apron deposited the grain at one side. The inventor received a prize from the Highland and Agricultural Society of Edinburgh, and pictures and full descriptions of his invention were published. Several models of this reaper were built in Great Britain, and it is said that four came to the United States; however this may be, Bell's machine was never generally adopted.

Soon afterward three men patented reapers in the United States: William Manning, Plainfield, New Jersey, 1831; Obed Hussey, Cincinnati, Ohio, 1833; and Cyrus Hall McCormick, Staunton, Virginia, 1834. Just how much they owed to Patrick Bell cannot be known, but it is probable that all had heard of his design if they had not seen his drawings or the machine itself. The first of these inventors, Manning of New Jersey, drops out of the story, for it is not known whether he ever made a machine other than his model. More persistent was Obed Hussey of Cincinnati, who soon moved to Baltimore to fight out the issue with McCormick. Hussey was an excellent mechanic. He patented several improvements to his machine and received high praise for the efficiency of the work. But he was soon outstripped in the race because he was weak in the essential qualities which made McCormick the greatest figure in the world of agricultural machinery. McCormick was more than a mechanic; he was a man of vision; and he had the enthusiasm of a crusader and superb genius for business organization and advertisement. His story has been told in another volume of this series.*

* *"The Age of Big Business", by Burton J. Hendrick.*

Though McCormick offered reapers for sale in 1834, he seems to have sold none in that year, nor any for six years afterwards. He sold two in 1840, seven in 1842, fifty in 1844. The machine was not really adapted to the hills of the Valley of Virginia, and farmers hesitated to buy a contrivance which needed the attention of a skilled mechanic. McCormick made a trip through the Middle West. In the rolling prairies, mile after mile of rich soil without a tree or a stone, he saw his future dominion. Hussey had moved East. McCormick did the opposite; he moved West, to Chicago, in 1847.

Chicago was then a town of hardly ten thousand, but McCormick foresaw its future, built a factory there, and manufactured five hundred machines for the harvest of 1848. From this time he went on from triumph to triumph. He formulated an elaborate business system. His machines were to be sold at a fixed price, payable in installments if desired, with a guarantee of satisfaction. He set up a system of agencies to give instruction or to supply spare parts. Advertising, chiefly by exhibitions and contests at fairs and other public gatherings, was another item of his programme. All would have failed, of course, if he had not built good machines, but he did build good machines, and was not daunted by the Government's refusal in 1848 to renew his original patent. He decided to make profits as a manufacturer rather than accept royalties as an inventor.

McCormick had many competitors, and some of them were in the field with improved devices ahead of him, but he always held his own, either by buying up the patent for a real improvement, or else by requiring his staff to invent something to do the same work. Numerous new devices to improve the harvester were patented, but the most important was an automatic attachment to bind the sheaves with wire. This was patented in 1872, and McCormick soon made it his own. The harvester seemed complete. One man drove the team, and the machine cut the grain, bound it in sheaves, and deposited them upon the ground.

Presently, however, complaints were heard of the wire tie. When the wheat was threshed, bits of wire got into the straw, and were swallowed by the cattle; or else the bits of metal got among the wheat itself and gave out sparks in grinding, setting some mills on fire. Two inventors, almost simultaneously, produced the remedy. Marquis L. Gorham, working for McCormick, and John F. Appleby, whose invention was purchased by William Deering, one of McCormick's chief competitors, invented binders which used twine. By 1880 the self-binding harvester was complete. No distinctive improvement has been made since, except to add strength and simplification. The machine now needed the services of only two men, one to drive and the other to shock the bundles, and could reap twenty acres or more a day, tie the grain into bundles of uniform size, and dump them in piles of five ready to be shocked.

Grain must be separated from the straw and chaff. The Biblical threshing floor, on which oxen or horses trampled out the grain, was still common in Washington's time, though it had been largely succeeded by the flail. In Great Britain several threshing machines were devised in the eighteenth century, but none was particularly successful. They were stationary, and it was necessary to bring the sheaves to them. The seventh patent issued by the United States, to Samuel Mulliken of Philadelphia, was for a threshing machine. The portable horse-power treadmill, invented in 1830 by Hiram A. and John A. Pitts of Winthrop, Maine, was presently coupled with a thresher, or "separator," and this outfit, with its men and horses, moving from farm to farm, soon became an autumn feature of every neighborhood. The treadmill was later on succeeded—by the traction engine, and the apparatus now in common use is an engine which draws the greatly improved threshing machine from farm to farm, and when the destination is reached, furnishes the power to drive the thresher. Many of these engines are adapted to the use of straw as fuel.

Another development was the combination harvester and thresher used on the larger farms of the West. This machine does not cut the wheat close to the ground, but the cutter-bar, over twenty-five feet in length, takes off the heads. The wheat is separated from the chaff and automatically weighed into sacks, which are dumped as fast as two expert sewers can work. The motive power is a traction engine or else twenty to thirty horses, and seventy-five acres a day can be reaped and threshed. Often another tractor pulling a dozen wagons follows and the sacks are picked up and hauled to the granary or elevator.

Haying was once the hardest work on the farm, and in no crop has machinery been more efficient. The basic idea in the reaper, the cutter-bar, is the whole of the mower, and the machine developed with the reaper. Previously Jeremiah Bailey, of Chester County, Pennsylvania, had patented in 1822 a machine drawn by horses carrying a revolving wheel with six scythes, which was widely used. The inventions of Manning, Hussey, and McCormick made the mower practicable. Hazard Knowles, an employee of the Patent Office, invented the hinged cutter-bar, which could be lifted over an obstruction, but never patented the invention. William F. Ketchum of Buffalo, New York, in 1844, patented the first machine intended to cut hay only, and dozens of others followed. The modern mowing machine was practically developed in the patent of Lewis Miller of Canton, Ohio, in 1858. Several times as many mowers as harvesters are sold, and for that matter, reapers without binding attachments are still manufactured.

Hayrakes and tedders seem to have developed almost of themselves. Diligent research has failed to discover any reliable information on the invention of the hayrake, though a horserake was patented as early as 1818. Joab Center of Hudson, New York, patented a machine for turning and spreading hay in 1834. Mechanical hayloaders have greatly reduced the amount of human labor. The hay-press makes storage and transportation easier and cheaper.

There are binders which cut and bind corn. An addition shocks the corn and deposits it upon the ground. The shredder and husker removes the ears, husks them, and shreds shucks, stalks, and fodder. Power shellers separate grain and cobs more than a hundred times as rapidly as a pair of human hands could do. One student of agriculture has estimated that it would require the whole agricultural population of the United States one hundred days to shell the average corn crop by hand, but this is an exaggeration.

The list of labor-saving machinery in agriculture is by no means exhausted. There are clover hullers, bean and pea threshers, ensilage cutters, manure spreaders, and dozens of others. On the dairy farm the cream separator both increases the quantity and improves the quality of the butter and saves time. Power also drives the churns. On many farms cows are milked and sheep are sheared by machines and eggs are hatched without hens.

There are, of course, thousands of farms in the country where machinery cannot be used to advantage and where the work is still done entirely or in part in the old ways.

Historians once were fond of marking off the story of the earth and of men upon the earth into distinct periods fixed by definite dates. One who attempts to look beneath the surface cannot accept this easy method of treatment. Beneath the surface new tendencies develop long before they demand recognition; an institution may be decaying long before its weakness is apparent. The American Revolution began not with the Stamp Act but at least a century earlier, as soon as the settlers realized that there were three thousand miles of sea between England and the rude country in which they found themselves; the Civil War began, if not in early Virginia, with the "Dutch Man of Warre that sold us twenty Negars," at least with Eli Whitney and his cotton gin.

Nevertheless, certain dates or short periods seem to be flowering times. Apparently all at once a flood of invention, a change of methods, a difference in organization, or a new psychology manifests itself. And the decade of the Civil War does serve as a landmark to mark the passing of one period in American life and the beginning of another; especially in agriculture; and as agriculture is the basic industry of the country it follows that with its mutations the whole superstructure is also changed.

The United States which fought the Civil War was vastly different from the United States which fronted the world at the close of the Revolution. The scant four million people of 1790 had grown to thirty-one and a half million. This growth had come chiefly by natural increase, but also by immigration, conquest, and annexation. Settlement had reached the Pacific Ocean, though there were great stretches of almost uninhabited territory between the settlements on the Pacific and those just beyond the Mississippi.

The cotton gin had turned the whole South toward the cultivation of cotton, though some States were better fitted for mixed farming, and their devotion to cotton meant loss in the end as subsequent events have proved. The South was not manufacturing any considerable proportion of the cotton it grew, but the textile industry was flourishing in New England. A whole series of machines similar to those used in Great Britain, but not identical, had been invented in America. American mills paid higher wages than British and in quantity production were far ahead of the British mills, in proportion to hands employed, which meant being ahead of the rest of the world.

Wages in America, measured by the world standard, were high, though as expressed in money, they seem low now. They were conditioned by the supply of free land, or land that was practically free. The wages paid were necessarily high enough to attract laborers from the soil which they might easily own if they chose. There was no fixed laboring class. The boy or girl in a textile mill often worked only a few years to save money, buy a farm, or to enter some business or profession.

The steamboat now, wherever there was navigable water, and the railroad, for a large part of the way, offered transportation to the boundless West. Steamboats traversed all the larger rivers and the lakes. The railroad was growing rapidly. Its lines had extended to more than thirty thousand miles. Construction went on during the war, and the transcontinental railway was in sight. The locomotive had approached standardization, and the American railway car was in form similar to that of the present day, though not so large, so comfortable, or so strong. The Pullman car, from which has developed the chair car, the dining car, and the whole list of special cars, was in process of development, and the automatic air brake of George Westinghouse was soon to follow.

Thus far had the nation progressed in invention and industry along the lines of peaceful development. But with the Civil War came a sudden and tremendous advance. No result of the Civil War, political or social, has more profoundly affected American life than the application to the farm, as a war necessity, of machinery on a great scale. So long as labor was plentiful and cheap, only a comparatively few farmers could be interested in expensive machinery, but when the war called the young men away the worried farmers gladly turned to the new machines and found that they were able not only to feed the Union, but also to export immense quantities of wheat to Europe, even during the war. Suddenly the West leaped into great prosperity. And long centuries of economic and social development were spanned within a few decades.

CHAPTER VI
AGENTS OF COMMUNICATION

Communication is one of man's primal needs. There was indeed a time when no formula of language existed, when men communicated with each other by means of gestures, grimaces, guttural sounds, or rude images of things seen; but it is impossible to conceive of a time when men had no means of communication at all. And at last, after long ages, men evolved in sound the names of the things they knew and the forms of speech; ages later, the alphabet and the art of writing; ages later still, those wonderful instruments of extension for the written and spoken word: the telegraph, the telephone, the modern printing press, the phonograph, the typewriter, and the camera.

The word "telegraph" is derived from Greek and means "to write far"; so it is a very exact word, for to write far is precisely what we do when we send a telegram. The word today, used as a noun, denotes the system of wires with stations and operators and messengers, girdling the earth and reaching into every civilized community, whereby news is carried swiftly by electricity. But the word was coined long before it was discovered that intelligence could be communicated by electricity. It denoted at first a system of semaphores, or tall poles with movable arms, and other signaling apparatus, set within sight of one another. There was such a telegraph line between Dover and London at the time of Waterloo; and this telegraph began relating the news of the battle, which had come to Dover by ship, to anxious London, when a fog set in and the Londoners had to wait until a courier on horseback arrived. And, in the very years when the real telegraph was coming into being, the United States Government, without a thought of electricity, was considering the advisability of setting up such a system of telegraphs in the United States.

The telegraph is one of America's gifts to the world. The honor for this invention falls to Samuel Finley Breese Morse, a New Englander of old Puritan stock. Nor is the glory that belongs to Morse in any way dimmed by the fact that he made use of the discoveries of other men who had been trying to unlock the secrets of electricity ever since Franklin's experiments. If Morse discovered no new principle, he is nevertheless the man of all the workers in electricity between his own day and Franklin's whom the world most delights to honor; and rightly so, for it is to such as Morse that the world is most indebted. Others knew; Morse saw and acted. Others had found out the facts, but Morse was the first to perceive the practical significance of those facts; the first to take steps to make them of service to his fellows; the first man of them all with the pluck and persistence to remain steadfast to his great design, through twelve long years of toil and privation, until his countrymen accepted his work and found it well done.

Morse was happy in his birth and early training. He was born in 1791, at Charlestown, Massachusetts. His father was a Congregational minister and a scholar of high standing, who, by careful management, was able to send his three sons to Yale College. Thither went young Samuel (or Finley, as he was called by his family) at the age of fourteen and came under the influence of Benjamin Silliman, Professor of Chemistry, and of Jeremiah Day, Professor of Natural Philosophy, afterwards President of Yale College, whose teaching gave him impulses which in later years led to the invention of the telegraph. "Mr. Day's lectures are very interesting," the young student wrote home in 1809; "they are upon electricity; he has given us some very fine experiments, the whole class taking hold of hands form the circuit of communication and we all receive the shock apparently at the same moment." Electricity, however, was only an alluring study. It afforded no means of livelihood, and Morse had gifts as an artist; in fact, he earned a part of his college expenses painting miniatures at five dollars apiece. He decided, therefore, that art should be his vocation.

A letter written years afterwards by Joseph M. Dulles of Philadelphia, who was at New Haven preparing for Yale when Morse was in his senior year, is worth reading here:

"I first became acquainted with him at New Haven, when about to graduate with the class of 1810, and had such an association as a boy preparing for college might have with a senior who was just finishing his course. Having come to New Haven under the care of Rev. Jedidiah Morse, the venerable father of the three Morses, all distinguished men, I was commended to the protection of Finley, as he was then commonly designated, and therefore saw him frequently during the brief period we were together. The father I regard as the gravest man I ever knew. He was a fine exemplar of the gentler type of the Puritan, courteous in manner, but stern in conduct and in aspect. He was a man of conflict, and a leader in the theological contests in New England in the early part of this century. Finley, on the contrary, bore the expression of gentleness entirely. In person rather above the ordinary height, well formed, graceful in demeanor, with a complexion, if I remember right, slightly ruddy, features duly proportioned, and often lightened with a genial and expressive smile. He was, altogether, a handsome young man, with manners unusually bland. It is needless to add that with intelligence, high culture, and general information, and with a strong bent to the fine arts, Mr. Morse was in 1810 an attractive young man. During the last year of his college life he occupied his leisure hours, with a view to his self-support, in taking the likenesses of his fellow-students on ivory, and no doubt with success, as he obtained afterward a very respectable rank as a portrait-painter. Many pieces of his skill were afterward executed in Charleston, South Carolina."*

* Prime, "The Life of Samuel F. B. Morse, LL.D.", p. 26.

That Morse was destined to be a painter seemed certain, and when, soon after graduating from Yale, he made the acquaintance of Washington Allston, an American artist of high standing, any doubts that may have existed in his mind as to his vocation were set at rest. Allston was then living in Boston, but was planning to return to England, where his name was well known, and it was arranged that young Morse should accompany him as his pupil. So in 1811 Morse went to England with Allston and returned to America four years later an accredited portrait painter, having studied not only under Allston but under the famous master, Benjamin West, and having met on intimate terms some of the great Englishmen of the time. He opened a studio in Boston, but as sitters were few, he made a trip through New England, taking commissions for portraits, and also visited Charleston, South Carolina, where some of his paintings may be seen today.

At Concord, New Hampshire, Morse met Miss Lucretia Walker, a beautiful and cultivated young woman, and they were married in 1818. Morse then settled in New York. His reputation as a painter increased steadily, though he gained little money, and in 1825 he was in Washington painting a portrait of the Marquis La Fayette, for the city of New York, when he heard from his father the bitter news of his wife's death in New Haven, then a journey of seven days from Washington. Leaving the portrait of La Fayette unfinished, the heartbroken artist made his way home.

Two years afterwards Morse was again obsessed with the marvels of electricity, as he had been in college. The occasion this time was a series of lectures on that subject given by James Freeman Dana before the New York Athenaeum in the chapel of Columbia College. Morse attended these lectures and formed with Dana an intimate acquaintance. Dana was in the habit of going to Morse's studio, where the two men would talk earnestly for long hours. But Morse was still devoted to his art; besides, he had himself and three children to support, and painting was his only source of income.

Back to Europe went Morse in 1829 to pursue his profession and perfect himself in it by three years' further study. Then came the crisis. Homeward bound on the ship Sully in the autumn of 1832, Morse fell into conversation with some scientific men who were on board. One of the passengers asked this question: "Is the velocity of electricity reduced by the length of its conducting wire?" To which his neighbor replied that electricity passes instantly over any known length of wire and referred to Franklin's experiments with several miles of wire, in which no appreciable time elapsed between a touch at one end and a spark at the other.

Here was a fact already well known. Morse must have known it himself. But the tremendous significance of that fact had never before occurred to him nor, so far as he knew, to any man. A recording telegraph! Why not? Intelligence delivered at one end of a wire instantly recorded at the other end, no matter how long the wire! It might reach across the continent or even round the earth. The idea set his mind on fire.

Home again in November, 1832, Morse found himself on the horns of a dilemma. To give up his profession meant that he would have no income; on the other hand, how could he continue wholeheartedly painting pictures while consumed with the idea of the telegraph? The idea would not down; yet he must live; and there were his three motherless children in New Haven. He would have to go on painting as well as he could and develop his telegraph in what time he could spare. His brothers, Richard and Sidney, were both living in New York and they did what they could for him, giving him a room in a building they had erected at Nassau and Beekman Streets. Morse's lot at this time was made all the harder by hopes raised and dashed to earth again. Congress had voted money for mural paintings for the rotunda of the Capitol. The artists were to be selected by a committee of which John Quincy Adams was chairman. Morse expected a commission for a part of the work, for his standing at that time was second to that of no American artist, save Allston, and Allston he knew had declined to paint any of the pictures and had spoken in his favor. Adams, however, as chairman of the committee was of the opinion that the pictures should be done by foreign artists, there being no Americans available, he thought, of sufficiently high standing to execute the work with fitting distinction. This opinion, publicly expressed, infuriated James Fenimore Cooper, Morse's friend, and Cooper wrote an attack on Adams in the New York Evening Post, but without signing it. Supposing Morse to be the author of this article, Adams summarily struck his name from the list of artists who were to be employed.

How very poor Morse was about this time is indicated by a story afterwards told by General Strother of Virginia, who was one of his pupils:

I engaged to become Morse's pupil and subsequently went to New York and found him in a room in University Place. He had three or four other pupils and I soon found that our professor had very little patronage.

I paid my fifty dollars for one-quarter's instruction. Morse was a faithful teacher and took as much interest in our progress as—more indeed than— we did ourselves. But he was very poor. I remember that, when my second quarter's pay was due, my remittance did not come as expected, and one day the professor came in and said, courteously: "Well Strother, my boy, how are we off for money?"

"Why professor," I answered, "I am sorry to say that I have been disappointed, but I expect a remittance next week."

"Next week," he repeated sadly, "I shall be dead by that time."

"Dead, sir?"

"Yes, dead by starvation."

I was distressed and astonished. I said hurriedly:

"Would ten dollars be of any service?"

"Ten dollars would save my life. That is all it would do."

I paid the money, all that I had, and we dined together. It was a modest meal, but good, and after he had finished, he said:

"This is my first meal for twenty-four hours. Strother, don't be an artist. It means beggary. Your life depends upon people who know nothing of your art and care nothing for you. A house dog lives better, and the very sensitiveness that stimulates an artist to work keeps him alive to suffering."*

Prime, p. 424.

In 1835 Morse received an appointment to the teaching staff of New York University and moved his workshop to a room in the University building in Washington Square. "There," says his biographer*, "he wrought through the year 1836, probably the darkest and longest year of his life, giving lessons to pupils in the art of painting while his mind was in the throes of the great invention." In that year he took into his confidence one of his colleagues in the University, Leonard D. Gale, who assisted him greatly, in improving the apparatus, while the inventor himself formulated the rudiments of the telegraphic alphabet, or Morse Code, as it is known today. At length all was ready for a test and the message flashed from transmitter to receiver. The telegraph was born, though only an infant as yet. "Yes, that room of the University was the birthplace of the Recording Telegraph," said Morse years later. On September 2, 1837, a successful experiment was made with seventeen hundred feet of copper wire coiled around the room, in the presence of Alfred Vail, a student, whose family owned the Speedwell Iron Works, at Morristown, New Jersey, and who at once took an interest in the invention and persuaded his father, Judge Stephen Vail, to advance money for experiments. Morse filed a petition for a patent in October and admitted his colleague Gale; as well as Alfred Vail, to partnership. Experiments followed at the Vail shops, all the partners working day and night in their enthusiasm. The apparatus was then brought to New York and gentlemen of the city were invited to the University to see it work before it left for Washington. The visitors were requested to write dispatches, and the words were sent round a three-mile coil of wire and read at the other end of the room by one who had no prior knowledge of the message.

Prime, p. 311.

In February, 1838, Morse set out for Washington with his apparatus, and stopped at Philadelphia on the invitation of the Franklin Institute to give a demonstration to a committee of that body. Arrived at Washington, he presented to Congress a petition, asking for an appropriation to enable him to build an experimental line. The question of the appropriation was referred to the Committee on Commerce, who reported favorably, and Morse then returned to New York to prepare to go abroad, as it was necessary for his rights that his invention should be patented in European countries before publication in the United States.

Morse sailed in May, 1838, and returned to New York by the steamship Great Western in April, 1839. His journey had not been very successful. He had found London in the excitement of the ceremonies of the coronation of Queen Victoria, and the British Attorney-General had refused him a patent on the ground that American newspapers had published his invention, making it public property. In France he had done better. But the most interesting result of the journey was something not related to the telegraph at all. In Paris he had met Daguerre, the celebrated Frenchman who had discovered a process of making pictures by sunlight, and Daguerre had given Morse the secret. This led to the first pictures taken by sunlight in the United States and to the first photographs of the human face taken anywhere. Daguerre had never attempted to photograph living objects and did not think it could be done, as rigidity of position was required for a long exposure. Morse, however, and his associate, John W. Draper, were very soon taking portraits successfully.

Meanwhile the affairs of the telegraph at Washington had not prospered. Congress had done nothing towards the grant which Morse had requested, notwithstanding the favorable report of its committee, and Morse was in desperate straits for money even to live on. He appealed to the Vails to assist him further, but they could not, since the panic of 1837 had impaired their resources. He earned small sums from his daguerreotypes and his teaching.

By December, 1842, Morse was in funds again; sufficiently, at least, to enable him to go to Washington for another appeal to Congress. And at last, on February 23, 1843, a bill appropriating thirty thousand dollars to lay the wires between Washington and Baltimore passed the House by a majority of six. Trembling with anxiety, Morse sat in the gallery of the House while the vote was taken and listened to the irreverent badinage of Congressmen as they discussed his bill. One member proposed an amendment to set aside half the amount for experiments in mesmerism, another suggested that the Millerites should have a part of the money, and so on; however, they passed the bill. And that night Morse wrote: "The long agony is over."

But the agony was not over. The bill had yet to pass the Senate. The last day of the expiring session of Congress arrived, March 3, 1843, and the Senate had not reached the bill. Says Morse's biographer:

In the gallery of the Senate Professor Morse had sat all the last day and evening of the session. At midnight the session would close. Assured by his friends that there was no possibility of the bill being reached, he left the Capitol and retired to his room at the hotel, dispirited, and well-nigh broken-hearted. As he came down to breakfast the next morning, a young lady entered, and, coming toward him with a smile, exclaimed:

"I have come to congratulate you!"

"For what, my dear friend?" asked the professor, of the young lady, who was Miss Annie G. Ellsworth, daughter of his friend the Commissioner of Patents.

"On the passage of your bill."

The professor assured her it was not possible, as he remained in the Senate-Chamber until nearly midnight, and it was not reached. She then informed him that her father was present until the close, and, in the last moments of the session, the bill was passed without debate or revision. Professor Morse was overcome by the intelligence, so joyful and unexpected, and gave at the moment to his young friend, the bearer of these good tidings, the promise that she should send the first message over the first line of telegraph that was opened.*

*Prime, p. 465.

Morse and his partners* then proceeded to the construction of the forty-mile line of wire between Baltimore and Washington. At this point

Ezra Cornell, afterwards a famous builder of telegraphs and founder of Cornell University, first appears in history as a young man of thirty-six. Cornell invented a machine to lay pipe underground to contain the wires and he was employed to carry out the work of construction. The work was commenced at Baltimore and was continued until experiment proved that the underground method would not do, and it was decided to string the wires on poles. Much time had been lost, but once the system of poles was adopted the work progressed rapidly, and by May, 1844, the line was completed. On the twenty-fourth of that month Morse sat before his instrument in the room of the Supreme Court at Washington. His friend Miss Ellsworth handed him the message which she had chosen: "WHAT HATH GOD WROUGHT!" Morse flashed it to Vail forty miles away in Baltimore, and Vail instantly flashed back the same momentous words, "WHAT HATH GOD WROUGHT!"

> ** The property in the invention was divided into sixteen*
> *shares (the partnership having been formed in 1838) of which*
> *Morse held 9, Francis O. J. Smith 4, Alfred Vail 2, Leonard*
> *D. Gale 2. In patents to be obtained in foreign countries,*
> *Morse was to hold 8 shares, Smith 5, Vail 2, Gale 1. Smith*
> *had been a member of Congress and Chairman of the Committee*
> *on Commerce. He was admitted to the partnership in*
> *consideration of his assisting Morse to arouse the interest*
> *of European Governments.*

Two days later the Democratic National Convention met in Baltimore to nominate a President and Vice-President. The leaders of the Convention desired to nominate Senator Silas Wright of New York, who was then in Washington, as running mate to James K. Polk, but they must know first whether Wright would consent to run as Vice-President. So they posted a messenger off to Washington but were persuaded at the same time to allow the new telegraph to try what it could do. The telegraph carried the offer to Wright and carried back to the Convention Wright's refusal of the honor. The delegates, however, would not believe the telegraph, until their own messenger, returning the next day, confirmed its message.

For a time the telegraph attracted little attention. But Cornell stretched the lines across the country, connecting city with city, and Morse and Vail improved the details of the mechanism and perfected the code. Others came after them and added further improvements. And it is gratifying to know that both Morse and Vail, as well as Cornell, lived to reap some return for their labor. Morse lived to see his telegraph span the continent, and link the New World with the Old, and died in 1872 full of honors.

Prompt communication of the written or spoken message is a demand even more insistent than prompt transportation of men and goods. By 1859 both the railroad and the telegraph had reached the old town of St. Joseph on the Missouri. Two thousand miles beyond, on the other side of plains and mountains and great rivers, lay prosperous California. The only transportation to California was by stage-coach, a sixty days' journey, or else across Panama, or else round the Horn, a choice of three evils. But to establish quicker communication, even though transportation might lag, the men of St. Joseph organized the Pony Express, to cover the great wild distance by riders on horseback, in ten or twelve days. Relay stations for the horses and men were set up at appropriate points all along the way, and a postboy dashed off from St. Joseph every twenty-four hours, on arrival of the train from the East. And for a time the Pony Express did its work and did it well. President Lincoln's First Inaugural was carried to California by the Pony Express; so was the news of the firing on Fort Sumter. But by 1869. the Pony Express was quietly superseded by the telegraph, which in that year had completed its circuits all the way to San Francisco, seven years ahead of the first transcontinental railroad. And in four more years Cyrus W. Field and Peter Cooper had carried to complete success the Atlantic Cable; and the Morse telegraph was sending intelligence across the sea, as well as from New York to the Golden Gate.

And today ships at sea and stations on land, separated by the sea, speak to one another in the language of the Morse Code, without the use of wires. Wireless, or radio, telegraphy was the invention of a nineteen-year-old boy, Guglielmo Marconi, an Italian; but it has been greatly extended and developed at the hands of four Americans: Fessenden, Alexanderson, Langmuir, and Lee De Forest. It was De Forest's invention that made possible transcontinental and transatlantic telephone service, both with and without wires.

The story of the telegraph's younger brother, and great ally in communication, the telephone of Alexander Graham Bell, is another pregnant romance of American invention. But that is a story by itself, and it begins in a later period and so falls within the scope of another volume of these Chronicles.*

* *"The Age of Big Business", by Burton J. Hendrick, "The Chronicle of America", vol. XXXIX.*

Wise newspapermen stiffened to attention when the telegraph began ticking. The New York Herald, the Sun, and the Tribune had been founded only recently and they represented a new type of journalism, swift, fearless, and energetic. The proprietors of these newspapers saw that this new instrument was bound to affect all newspaperdom profoundly. How was the newspaper to cope with the situation and make use of the news that was coming in and would be coming in more and more over the wires?

For one thing, the newspapers needed better printing machinery. The application of steam, or any mechanical power, to printing in America was only begun. It had been introduced by Robert Hoe in the very years when Morse was struggling to perfect the telegraph. Before that time newspapers were printed in the United States, on presses operated as Franklin's press had been operated, by hand. The New York Sun, the pioneer of cheap modern newspapers, was printed by hand in 1833, and four hundred impressions an hour was the highest speed of one press. There had been, it is true, some improvements over Franklin's printing press. The Columbian press of George Clymer of Philadelphia, invented in 1816, was a step forward. The Washington press, patented in 1829 by Samuel Rust of New York, was another step forward. Then had come Robert Hoe's double-cylinder, steamdriven printing press. But a swifter machine was wanted. And so in 1845 Richard March Hoe, a son of Robert Hoe, invented the revolving or rotary press, on the principle of which larger and larger machines have been built—machines so complex and wonderful that they baffle description; which take in reels of white paper and turn out great newspapers complete, folded and counted, at the rate of a hundred thousand copies an hour. American printing machines are in use today the world over. The London Times is printed on American machines.

Hundreds of new inventions and improvements on old inventions followed hard on the growth of the newspaper, until it seemed that the last word had been spoken. The newspapers had the wonderful Hoe presses; they had cheap paper; they had excellent type, cast by machinery; they had a satisfactory process of multiplying forms of type by stereotyping; and at length came a new process of making pictures by photo-engraving, supplanting the old-fashioned process of engraving on wood. Meanwhile, however, in one important department of the work, the newspapers had made no advance whatever. The newspapers of New York in the year 1885, and later, set up their type by the same method that Benjamin Franklin used to set up the type for The Pennsylvania Gazette. The compositor stood or sat at his "case," with his "copy" before him, and picked the type up letter by letter until he had filled and correctly spaced a line. Then he would set another line, and so on, all with his hands. After the job was completed, the type had to be distributed again, letter by letter. Typesetting was slow and expensive.

This labor of typesetting was at last generally done away with by the invention of two intricate and ingenious machines. The linotype, the invention of Ottmar Mergenthaler of Baltimore, came first; then the monotype of Tolbert Lanston, a native of Ohio. The linotype is the favorite composing machine for newspapers and is also widely used in typesetting for books, though the monotype is preferred by book printers. One or other of these machines has today replaced, for the most part, the old hand compositors in every large printing establishment in the United States.

While the machinery of the great newspapers was being developed, another instrument of communication, more humble but hardly less important in modern life, was coming into existence. The typewriter is today in every business office and is another of America's gifts to the commercial world. One might attempt to trace the typewriter back to the early seals, or to the name plates of the Middle Ages, or to the records of the British Patent Office, for 1714, which mention a machine for embossing. But it would be difficult to establish the identity of these contrivances with the modern typewriter.

Two American devices, one of William Burt in 1829, for a "typographer," and another of Charles Thurber, of Worcester, Massachusetts, in 1843, may also be passed over. Alfred Ely Beach made a model for a typewriter as early as 1847, but neglected it for other things, and his next effort in printing machines was a device for embossing letters for the blind. His typewriter had many of the features of the modern typewriter, but lacked a satisfactory method of inking the types. This was furnished by S. W. Francis of New York, whose machine, in 1857, bore a ribbon saturated with ink. None of these machines, however, was a commercial success. They were regarded merely as the toys of ingenious men.

The accredited father of the typewriter was a Wisconsin newspaperman, Christopher Latham Sholes, editor, politician, and anti-slavery agitator. A strike of his printers led him to unsuccessful attempts to invent a typesetting machine. He did succeed, however, in making, in collaboration with another printer, Samuel W. Soule, a numbering machine, and a friend, Carlos Glidden, to whom this ingenious contrivance was shown, suggested a machine to print letters.

The three friends decided to try. None had studied the efforts of previous experimenters, and they made many errors which might have been avoided. Gradually, however, the invention took form. Patents were obtained in June, 1868, and again in July of the same year, but the machine was neither strong nor trustworthy. Now appeared James Densmore and bought a share in the machine, while Soule and Glidden retired. Densmore furnished the funds to build about thirty models in succession, each a little better than the preceding. The improved machine was patented in 1871, and the partners felt that they were ready to begin manufacturing.

Wisely they determined, in 1873, to offer their machine to Eliphalet Remington and Sons, then manufacturing firearms, sewing machines, and the like, at Ilion, New York. Here, in well-equipped machine shops it was tested, strengthened, and improved. The Remingtons believed they saw a demand for the machine and offered to buy the patents, paying either a lump sum, or a royalty. It is said that Sholes preferred the ready cash and received twelve thousand dollars, while Densmore chose the royalty and received a million and a half.

The telegraph, the press, and the typewriter are agents of communication for the written word. The telephone is an agent for the spoken word. And there is another instrument for recording sound and reproducing it, which should not be forgotten. It was in 1877 that Thomas Alva Edison completed the first phonograph. The air vibrations set up by the human voice were utilized to make minute indentations on a sheet of tinfoil placed over a metallic cylinder, and the machine would then reproduce the sounds which had caused the indentations. The record wore out after a few reproductions, however, and Edison was too busy to develop his idea further for a time, though later he returned to it.

The phonograph today appears under various names, but by whatever name they are called, the best machines reproduce with wonderful fidelity the human voice, in speech or song, and the tones of either a single instrument or a whole orchestra. The most distinguished musicians are glad to do their best for the preservation and reproduction of their art, and through these machines, good music is brought to thousands to whom it could come in no other way.

The camera bears a large part in the diffusion of intelligence, and the last half century in the United States has seen a great development in photography and photoengraving. The earliest experiments in photography belong almost exclusively to Europe. Morse, as we have seen, introduced the secret to America and interested his friend John W. Draper, who had a part in the perfection of the dry plate and who was one of the first, if not the first, to take a portrait by photography.

The world's greatest inventor in photography is, however, George Eastman, of Rochester. It was in 1888 that Eastman introduced a new camera, which he called by the distinctive name Kodak, and with it the slogan: "You press the button, we do the rest." This first kodak was loaded with a roll of sensitized paper long enough for a hundred exposures. Sent to the makers, the roll could itself be developed and pictures could be printed from it. Eastman had been an amateur photographer when the fancy was both expensive and tedious. Inventing a method of making dry plates, he began to manufacture them in a small way as early as 1880. After the first kodak, there came others filled with rolls of sensitized nitro-cellulose film. Priority in the invention of the cellulose film, instead of glass, which has revolutionized photography, has been decided by the courts to belong to the Reverend Hannibal Goodwin, but the honor none the less belongs to Eastman, who independently worked out his process and gave photography to the millions. The introduction by the Eastman Kodak Company of a film cartridge which could be inserted or removed without retiring to a dark room removed the chief difficulty in the way of amateurs, and a camera of some sort, varying in price from a dollar or two to as many hundreds, is today an indispensable part of a vacation equipment.

In the development of the animated pictures Thomas Alva Edison has played a large part. Many were the efforts to give the appearance of movement to pictures before the first real entertainment was staged by Henry Heyl of Philadelphia. Heyl's pictures were on glass plates fixed in the circumference of a wheel, and each was brought and held for a part of a second before the lens. This method was obviously too slow and too expensive. Edison with his keen mind approached the difficulty and after a prolonged series of experiments arrived at the decision that a continuous tape-like film would be necessary. He invented the first practical "taking" camera and evoked the enthusiastic cooperation of George Eastman in the production of this tape-like film, and the modern motion picture was born. The projecting machine was substantially like the "taking" camera and was so used. Other inventors, such as Paul in England and Lumiere in France, produced other types of projecting machines, which differed only in mechanical details.

When the motion picture was taken up in earnest in the United States, the world stared in astonishment at the apparent recklessness of the early managers. The public responded, however, and there is hardly a hamlet in the nation where there is not at least one moving-picture house. The most popular actors have been drawn from the speaking stage into the "movies," and many new actors have been developed. In the small town, the picture theater is often a converted storeroom, but in the cities, some of the largest and most attractive theaters have been given over to the pictures, and others even more luxurious have been specially built. The Eastman Company alone manufactures about ten thousand miles of film every month.

Besides affording amusement to millions, the moving picture has been turned to instruction. Important news events are shown on the screen, and historical events are preserved for posterity by depositing the films in a vault. What would the historical student not give for a film faithfully portraying the inauguration of George Washington! The motion picture has become an important factor in instruction in history and science in the schools and this development is still in its infancy.

CHAPTER VII
THE STORY OF RUBBER

One day in 1852, at Trenton, New Jersey, there appeared in the Circuit Court of the United States two men, the legal giants of their day, to argue the case of Goodyear vs. Day for infringement of patent. Rufus Choate represented the defendant and Daniel Webster the plaintiff. Webster, in the course of his plea, one of the most brilliant and moving ever uttered by him, paused for a moment, drew from himself the attention of those who were hanging upon his words, and pointed to his client. He would have them look at the man whose cause he pleaded: a man of fifty-two, who looked fifteen years older, sallow, emaciated from disease, due to long privations, bitter disappointments, and wrongs. This was Charles Goodyear, inventor of the process which put rubber into the service of the world. Said Webster:

"And now is Charles Goodyear the discoverer of this invention of vulcanized rubber? Is he the first man upon whose mind the idea ever flashed, or to whose intelligence the fact ever was disclosed, that by carrying heat to a certain height it would cease to render plastic the India Rubber and begin to harden and metallize it? Is there a man in the world who found out that fact before Charles Goodyear? Who is he? Where is he? On what continent does he live? Who has heard of him? What books treat of him? What man among all the men on earth has seen him, known him, or named him? Yet it is certain that this discovery has been made. It is certain that it exists. It is certain that it is now a matter of common knowledge all over the civilized world. It is certain that ten or twelve years ago it was not knowledge. It is certain that this curious result has grown into knowledge by somebody's discovery and invention. And who is that somebody? The question was put to my learned opponent by my learned associate. If Charles Goodyear did not make this discovery, who did make it? Who did make it? Why, if our learned opponent had said he should endeavor to prove that some one other than Mr. Goodyear had made this discovery, that would have been very fair. I think the learned gentleman was very wise in not doing so. For I have thought often, in the course of my practice in law, that it was not very advisable to raise a spirit that one could not conveniently lay again. Now who made this discovery? And would it not be proper? I am sure it would. And would it not be manly? I am sure it would. Would not my learned friend and his coadjutor have acted a more noble part, if they had stood up and said that this invention was not Goodyear's, but it was an invention of such and such a man, in this or that country? On the contrary they do not meet Goodyear's claim by setting up a distinct claim of anybody else. They attempt to prove that he was not the inventor by little shreds and patches of testimony. Here a little bit of sulphur, and there a little parcel of lead; here a little degree of heat, a little hotter than would warm a man's hands, and in which a man could live for ten minutes or a quarter of an hour; and yet they never seem to come to the point. I think it is because their materials did not allow them to come to the manly assertion that somebody else did make this invention, giving to that somebody a local habitation and a name. We want to know the name, and the habitation, and the location of the man upon the face of this globe, who invented vulcanized rubber, if it be not he, who now sits before us.

"Well there are birds which fly in the air, seldom lighting, but often hovering. Now I think this is a question not to be hovered over, not to be brooded over, and not to be dealt with as an infinitesimal quantity of small things. It is a case calling for a manly admission and a manly defense. I ask again, if there is anybody else than Goodyear who made this invention, who is he? Is the discovery so plain that it might have come about by accident? It is likely to work important changes in the arts everywhere. IT INTRODUCES QUITE A NEW MATERIAL INTO THE MANUFACTURE OF THE ARTS, THAT MATERIAL BEING NOTHING LESS THAN ELASTIC METAL. It is hard like metal and as elastic as pure original gum elastic. Why, that is as great and momentous a phenomenon occurring to men in the progress of their knowledge, as it would be for a man to show that iron and gold could remain iron and gold and yet become elastic like India Rubber. It would be just such another result. Now, this fact cannot be denied; it cannot be secreted; it cannot be kept out of sight; somebody has made this invention. That is certain. Who is he? Mr. Hancock has been referred to. But he expressly acknowledges Goodyear to be the first inventor. I say that there is not in the world a human being that can stand up and say that it is his invention, except the man who is sitting at that table."

The court found for the plaintiff, and this decision established for all time the claim of the American, Charles Goodyear, to be the sole inventor of vulcanized rubber.

This trial may be said to be the dramatic climax in the story of rubber. It celebrated the hour when the science of invention turned a raw product — which had tantalized by its promise and wrought ruin by its treachery — into a manufacture adaptable to a thousand uses, adding to man's ease and health and to the locomotion, construction, and communication of modern life.

When Columbus revisited Hayti on his second voyage, he observed some natives playing with a ball. Now, ball games are the oldest sport known. From the beginning of his history man, like the kitten and the puppy, has delighted to play with the round thing that rolls. The men who came with Columbus to conquer the Indies had brought their Castilian wind-balls to play with in idle hours. But at once they found that the balls of Hayti were incomparably superior toys; they bounced better. These high bouncing balls were made, so they learned, from a milky fluid of the consistency of honey which the natives procured by tapping certain trees and then cured over the smoke of palm nuts. A discovery which improved the delights of ball games was noteworthy.

The old Spanish historian, Herrera, gravely transcribed in his pages all that the governors of Hayti reported about the bouncing balls. Some fifty years later another Spanish historian related that the natives of the Amazon valley made shoes of this gum; and that Spanish soldiers spread their cloaks with it to keep out the rain. Many years later still, in 1736, a French astronomer, who was sent by his government to Peru to measure an arc of the meridian, brought home samples of the gum and reported that the natives make lights of it, "which burn without a wick and are very bright," and "shoes of it which are waterproof, and when smoked they have the appearance of leather. They also make pear-shaped bottles on the necks of which they fasten wooden tubes. Pressure on the bottle sends the liquid squirting out of the tube, so they resemble syringes." Their name for the fluid, he added, was "cachuchu" — caoutchouc, we now write it. Evidently the samples filled no important need at the time, for we hear no more of the gum until thirty-four years afterward. Then, so an English writer tells us, a use was found for the gum — and a name. A stationer accidentally discovered that it would erase pencil marks, And, as it came from the Indies and rubbed, of course it was "India rubber."

About the year 1820 American merchantmen, plying between Brazil and New England, sometimes carried rubber as ballast on the home voyage and dumped it on the wharves at Boston. One of the shipmasters exhibited to his friends a pair of native shoes fancifully gilded. Another, with more foresight, brought home five hundred pairs, ungilded, and offered them for sale. They were thick, clumsily shaped, and heavy, but they sold. There was a demand for more. In a few years half a million pairs were being imported annually. New England manufacturers bid against one another along the wharves for the gum which had been used as ballast and began to make rubber shoes.

European vessels had also carried rubber home; and experiments were being made with it in France and Britain. A Frenchman manufactured suspenders by cutting a native bottle into fine threads and running them through a narrow cloth web. And Macintosh, a chemist of Glasgow, inserted rubber treated with naphtha between thin pieces of cloth and evolved the garment that still bears his name.

At first the new business in rubber yielded profits. The cost of the raw material was infinitesimal; and there was a demand for the finished articles. In Roxbury, Massachusetts, a firm manufacturing patent leather treated raw rubber with turpentine and lampblack and spread it on cloth, in an effort to produce a waterproof leather. The process appeared to be a complete success, and a large capital was employed to make handsome shoes and clothing out of the new product and in opening shops in the large cities for their sale. Merchants throughout the country placed orders for these goods, which, as it happened, were made and shipped in winter.

But, when summer came, the huge profits of the manufacturers literally melted away, for the beautiful garments decomposed in the heat; and loads of them, melting and running together, were being returned to the factory. And they filled Roxbury with such noisome odors that they had to be taken out at dead of night and buried deep in the earth.

And not only did these rubber garments melt in the heat. It presently transpired that severe frost stiffened them to the rigidity of granite. Daniel Webster had had some experience in this matter himself. "A friend in New York," he said, "sent me a very fine cloak of India Rubber, and a hat of the same material. I did not succeed very well with them. I took the cloak one day and set it out in the cold. It stood very well by itself. I surmounted it with the hat, and many persons passing by supposed they saw, standing by the porch, the Farmer of Marshfield."

It was in the year 1834, shortly after the Roxbury manufacturers had come to realize that their process was worthless and that their great fortune was only a mirage, and just before these facts became generally known, that Charles Goodyear made his entrance on the scene. He appeared first as a customer in the company's store in New York and bought a rubber life-preserver. When he returned some weeks later with a plan for improving the tube, the manager confided to him the sad tragedy of rubber, pointing out that no improvement in the manufactured articles would meet the difficulty, but that fame and fortune awaited the inventor of a process that would keep rubber dry and firm and flexible in all weathers.

Goodyear felt that he had a call from God. "He who directs the operations of the mind," he wrote at a later date, "can turn it to the development of the properties of Nature in his own way, and at the time when they are specially needed. The creature imagines he is executing some plan of his own, while he is simply an instrument in the hands of his Maker for executing the divine purposes of beneficence to the race." It was in the spirit of a crusader, consecrated to a particular service, that this man took up the problem of rubber. The words quoted are a fitting preface for the story of the years that followed, which is a tale of endurance and persistent activity under sufferings and disappointments such as are scarcely paralleled even in the pages of invention, darkened as they often are by poverty and defeat.

Charles Goodyear was born at New Haven, December 29, 1800, the son of Amasa Goodyear and descendant of Stephen Goodyear who was associated with Theophilus Eaton, the first governor of the Puritan colony of New Haven. It was natural that Charles should turn his mind to invention, as he did even when a boy; for his father, a pioneer in the manufacture of American hardware, was the inventor of a steel hayfork which replaced the heavy iron fork of prior days and lightened and expedited the labor of the fields. When Charles was seven his father moved to Naugatuck and manufactured the first pearl buttons made in America; during the War of 1812 the Goodyear factory supplied metal buttons to the Government. Charles, a studious, serious boy, was

*the close companion of his father. His deeply religious nature
manifested itself early, and he joined the Congregational Church
when he was sixteen. It was at first his intention to enter the
ministry, which seemed to him to offer the most useful career
of service, but, changing his mind, he went to Philadelphia to
learn the hardware business and on coming of age was admitted
to partnership in a firm established there by his father. The firm
prospered for a time, but an injudicious extension of credit led
to its suspension. So it happened that Goodyear in 1834, when
he became interested in rubber, was an insolvent debtor, liable,
under the laws of the time, to imprisonment. Soon afterward,
indeed, he was lodged in the*

Debtor's Prison in Philadelphia.

*It would seem an inauspicious hour to begin a search which
might lead him on in poverty for years and end nowhere. But,
having seen the need for perfect rubber, the thought had come to
him, with the force of a religious conviction, that "an object so
desirable and so important, and so necessary to man's comfort, as
the making of gum-elastic available to his use, was most certainly
placed within his reach." Thereafter he never doubted that God
had called him to this task and that his efforts would be crowned
with success. Concerning his prison experiences, of which the
first was not to be the last, he says that "notwithstanding the
mortification attending such a trial," if the prisoner has a real
aim "for which to live and hope over he may add firmness to
hope, and derive lasting advantage by having proved to himself
that, with a clear conscience and a high purpose, a man may
be as happy within prison walls as in any other (even the most
fortunate) circumstances in life." With this spirit he met every
reverse throughout the ten hard years that followed.* Luckily,
as he says, his first experiments required no expensive
equipment. Fingers were the best tools for working the
gum. The prison officials allowed him a bench and a
marble slab, a friend procured him a few dollars' worth of

gum, which sold then at five cents a pound, and his wife contributed her rolling pin. That was the beginning.

For a time he believed that, by mixing the raw gum with magnesia and boiling it in lime, he had overcome the stickiness which was the inherent difficulty. He made some sheets of white rubber which were exhibited, and also some articles for sale. His hopes were dashed when he found that weak acid, such as apple juice or vinegar, destroyed his new product. Then in 1836 he found that the application of aqua fortis, or nitric acid, produced a "curing" effect on the rubber and thought that he had discovered the secret. Finding a partner with capital, he leased an abandoned rubber factory on Staten Island. But his partner's fortune was swept away in the panic of 1837, leaving Goodyear again an insolvent debtor. Later he found another partner and went to manufacturing in the deserted plant at Roxbury, with an order from the Government for a large number of mail bags. This order was given wide publicity and it aroused the interest of manufacturers throughout the country. But by the time the goods were ready for delivery the first bags made had rotted from their handles. Only the surface of the rubber had been "cured."

This failure was the last straw, as far as Goodyear's friends were concerned. Only his patient and devoted wife stood by him; she had labored, known want, seen her children go hungry to school, but she seems never to have reproached her husband nor to have doubted his ultimate success. The gentleness and tenderness of his deportment in the home made his family cling to him with deep affection and bear willingly any sacrifice for his sake; though his successive failures generally meant a return of the inventor to the debtor's prison and the casting of his family upon charity.

The nitric acid process had not solved the problem but it had been a real step forward. It was in the year 1839, by an accident, that he discovered the true process of vulcanization which cured not the surface alone but the whole mass. He was trying to harden the gum by boiling it with sulphur on his wife's cookstove when he let fall a lump of it on the red hot iron top. It vulcanized instantly. This was an accident which only Goodyear could have interpreted. And it was the last. The strange substance from the jungles of the tropics had been mastered. It remained, however, to perfect the process, to ascertain the accurate formula and the exact degree of heat. The Goodyears were so poor during these years that they received at any time a barrel of flour from a neighbor thankfully. There is a tradition that on one occasion, when Goodyear desired to cross between Staten Island and New York, he had to give his umbrella to the ferry master as security for his fare, and that the name of the ferry master was Cornelius Vanderbilt, "a man who made much money because he took few chances." The incident may easily have occurred, though the ferry master could hardly have been Vanderbilt himself, unless it had been at an earlier date. Another tradition says that one of Goodyear's neighbors described him to an inquisitive stranger thus: "You will know him when you see him; he has on an India rubber cap, stock, coat, vest, and shoes, and an India rubber purse WITHOUT A CENT IN IT!"

Goodyear's trials were only beginning. He had the secret at last, but nobody would believe him. He had worn out even the most sanguine of his friends. "That such indifference to this discovery, and many incidents attending it, could have existed in an intelligent and benevolent community," wrote Goodyear later, "can only be accounted for by existing circumstances in that community The great losses that had been sustained in the manufacture of gum-elastic: the length of time the inventor had spent in what appeared to them to be entirely fruitless efforts to accomplish anything with it; added to his recent misfortunes and disappointments, all conspired, with his utter destitution, to produce a state of things as unfavorable to the promulgation of the discovery as can well be imagined. He, however, felt in duty bound to beg in earnest, if need be, sooner than that the discovery should be lost to the world and to himself.... How he subsisted at this period charity alone can tell, for it is as well to call things by their right names; and it is little else than charity when the lender looks upon what he parts with as a gift. The pawning or selling some relic of better days or some article of necessity was a frequent expedient. His library had long since disappeared, but shortly after the discovery of this process, he collected and sold at auction the schoolbooks of his children, which brought him the trifling sum of five dollars; small as the amount was, it enabled him to proceed. At this step he did not hesitate. The occasion, and the certainty of success, warranted the measure which, in other circumstances, would have been sacrilege."

His itinerary during those years is eloquent. Wherever there was a man, who had either a grain of faith in rubber or a little charity for a frail and penniless monomaniac, thither Goodyear made his way. The goal might be an attic room or shed to live in rent free, or a few dollars for a barrel of flour for the family and a barrel of rubber for himself, or permission to use a factory's ovens after hours and to hang his rubber over the steam valves while work went on. From Woburn in 1839, the year of his great discovery, he went to Lynn, from Lynn back to the deserted factory at Roxbury. Again to Woburn, to Boston, to Northampton, to Springfield, to Naugatuck; in five years as many removes. When he lacked boat or railway fare, and he generally did, he walked through winds and rains and drifting snow, begging shelter at some cottage or farm where a window lamp gleamed kindly.

Goodyear took out his patent in 1844. The process he invented has been changed little, if at all, from that day to this. He also invented the perfect India rubber cloth by mixing fiber with the gum a discovery he considered rightly as secondary in importance only to vulcanization. When he died in 1860 he had taken out sixty patents on rubber manufactures. He had seen his invention applied to several hundred uses, giving employment to sixty thousand persons, producing annually eight million dollars' worth of merchandise—numbers which would form but a fraction of the rubber statistics of today.

Everybody, the whole civilized world round, uses rubber in one form or another. And rubber makes a belt around the world in its natural as well as in its manufactured form. The rubber-bearing zone winds north and south of the equator through both hemispheres. In South America rubber is the latex of certain trees, in Africa of trees and vines. The best "wild" rubber still comes from Para in Brazil. It is gathered and prepared for shipment there today by the same methods the natives used four hundred years ago. The natives in their canoes follow the watercourses into the jungles. They cut V-shaped or spiral incisions in the trunks of the trees that grow sheer to sixty feet before spreading their shade. At the base of the incisions they affix small clay cups, like swallows' nests. Over the route they return later with large gourds in which they collect the fluid from the clay cups. The filled gourds they carry to their village of grass huts and there they build their smoky fires of oily palm nuts. Dipping paddles into the fluid gum they turn and harden it, a coating at a time, in the smoke. The rubber "biscuit" is cut from the paddle with a wet knife when the desired thickness has been attained.

Goodyear lived for sixteen years after his discovery of the vulcanization process. During the last six he was unable to walk without crutches. He was indifferent to money. To make his discoveries of still greater service to mankind was his whole aim. It was others who made fortunes out of his inventions. Goodyear died a poor man.

In his book, a copy of which was printed on gumelastic sheets and bound in hard rubber carved, he summed up his philosophy in this statement: "The writer is not disposed to repine and say that he has planted and others have gathered the fruits. The advantages of a career in life should not be estimated exclusively by the standard of dollars and cents, as it is too often done. Man has just cause for regret when he sows and no one reaps."

CHAPTER VIII
PIONEERS OF THE MACHINE SHOP

There is a tinge of melancholy about the life of such a pioneer as Oliver Evans, that early American mechanic of great genius, whose story is briefly outlined in a preceding chapter. Here was a man of imagination and sensibility, as well as practical power; conferring great benefits on his countrymen, yet in chronic poverty; derided by his neighbors, robbed by his beneficiaries; his property, the fruit of his brain and toil, in the end malevolently destroyed. The lot of the man who sees far ahead of his time, and endeavors to lead his fellows in ways for which they are not prepared, has always been hard.

John Stevens, too, as we have seen, met defeat when he tried to thrust a steam railroad on a country that was not yet ready for it. His mechanical conceptions were not marked by genius equal to that of Evans, but they were still too far advanced to be popular. The career of Stevens, however, presents a remarkable contrast to that of Evans in other respects. Evans was born poor (in Delaware, 1755) and remained poor all his life. Stevens was born rich (in New York City, 1749) and remained rich all his life. Of the family of Evans nothing is known either before or after him. Stevens, on the contrary, belonged to one of the best known and most powerful families in America. His grandfather, John Stevens I, came from England in 1699 and made himself a lawyer and a great landowner. His father, John Stevens II, was a member from New Jersey of the Continental Congress and presided at the New Jersey Convention which ratified the Constitution.

John Stevens III was graduated at King's College (Columbia) in 1768. He held public offices during the Revolution. To him, perhaps more than to any other man, is due the Patent Act of 1790, for the protection of American inventors, for that law was the result of a petition which he made to Congress and which, being referred to a committee, was favorably reported. Thus we may regard John Stevens as the father of the American patent law.

John Stevens owned the old Dutch farm on the Hudson on which the city of Hoboken now stands. The place had been in possession of the Bayard family, but William Bayard, who lived there at the time of the Revolution, was a Loyalist, and his house on Castle Point was burned down and his estate confiscated. After the Revolution Stevens acquired the property. He laid it out as a town in 1804, made it his summer residence, and established there the machine shops in which he and his sons carried on their mechanical experiments.

These shops were easily the largest and bestequipped in the Union when in 1838 John Stevens died at the age of ninety. The four brothers, John Cox, Robert Livingston, James Alexander, and Edwin Augustus, worked harmoniously together. "No one ever heard of any quarrel or dissension in the Stevens family. They were workmen themselves, and they were superior to their subordinates because they were better engineers and better men of business than any other folk who up to that time had undertaken the business of transportation in the United States."*

Abram S. Hewitt. Quoted in Iles, "Leading American Inventors", p. 37.

The youngest of these brothers, Edwin Augustus Stevens, dying in 1868, left a large part of his fortune to found the Stevens Institute of Technology, afterwards erected at Hoboken not far from the old family homestead on Castle Point. The mechanical star of the family, however, was the second brother, Robert Livingston Stevens, whose many inventions made for the great improvement of transportation both by land and water. For a quarter of a century, from 1815 to 1840, he was the foremost builder of steamboats in America, and under his hand the steamboat increased amazingly in speed and efficiency. He made great contributions to the railway. The first locomotives ran upon wooden stringers plated with strap iron. A loose end—"a snakehead" it was called—sometimes curled up and pierced through the floor of a car, causing a wreck. The solid metal T-rail, now in universal use, was designed by Stevens and was first used on the Camden and Amboy Railroad, of which he was president and his brother Edwin treasurer and manager. The swivel truck and the cow-catcher, the modern method of attaching rails to ties, the vestibule car, and many improvements in the locomotive were also first introduced on the Stevens road.

The Stevens brothers exerted their influence also on naval construction. A double invention of Robert and Edwin, the forced draft, to augment steam power and save coal, and the air-tight fireroom, which they applied to their own vessels, was afterwards adopted by all navies. Robert designed and projected an ironclad battleship, the first one in the world. This vessel, called the Stevens Battery, was begun by authority of the Government in 1842; but, owing to changes in the design and inadequate appropriations by Congress, it was never launched. It lay for many years in the basin at

Hoboken an unfinished hulk. Robert died in 1856. On the outbreak of the Civil War, Edwin tried to revive the interest of the Government, but by that time the design of the Stevens Battery was obsolete, and Edwin Stevens was an old man. So the honors for the construction of the first ironclad man-of-war to fight and win a battle went to John Ericsson, that other great inventor, who built the famous Monitor for the Union Government.

Carlyle's oft-quoted term, "Captains of Industry," may fittingly be applied to the Stevens family. Strong, masterful, and farseeing, they used ideas, their own and those of others, in a large way, and were able to succeed where more timorous inventors failed. Without the stimulus of poverty they achieved success, making in their shops that combination of men and material which not only added to their own fortunes but also served the world.

We left Eli Whitney defeated in his efforts to divert to himself some adequate share of the untold riches arising from his great invention of the cotton gin. Whitney, however, had other sources of profit in his own character and mechanical ability. As early as 1798 he had turned his talents to the manufacture of firearms. He had established his shops at Whitneyville, near New Haven; and it was there that he worked out another achievement quite as important economically as the cotton gin, even though the immediate consequences were less spectacular: namely, the principle of standardization or interchangeability in manufacture.

This principle is the very foundation today of all American large-scale production. The manufacturer produces separately thousands of copies of every part of a complicated machine, confident that an equal number of the complete machine will be assembled and set in motion. The owner of a motor car, a reaper, a tractor, or a sewing machine, orders, perhaps by telegraph or telephone, a broken or lost part, taking it for granted that the new part can be fitted easily and precisely into the place of the old.

Though it is probable that this idea of standardization, or interchangeability, originated independently in Whitney's mind, and though it is certain that he and one of his neighbors, who will be mentioned presently, were the first manufacturers in the world to carry it out successfully in practice, yet it must be noted that the idea was not entirely new. We are told that the system was already in operation in England in the manufacture of ship's blocks. From no less an authority than Thomas Jefferson we learn that a French mechanic had previously conceived the same idea.* But, as no general result whatever came from the idea in either France or England, the honors go to Whitney and North, since they carried it to such complete success that it spread to other branches of manufacturing.

And in the face of opposition. When Whitney wrote that his leading object was "to substitute correct and effective operations of machinery for that skill of the artist which is acquired only by long practice and experience," in order to make the same parts of different guns "as much like each other as the successive impressions of a copper-plate engraving," he was laughed to scorn by the ordnance officers of France and England. "Even the Washington officials," says Roe, "were sceptical and became uneasy at advancing so much money without a single gun having been completed, and Whitney went to Washington, taking with him ten pieces of each part of a musket. He exhibited these to the Secretary of War and the army officers interested, as a succession of piles of different parts. Selecting indiscriminately from each of the piles, he put together ten muskets, an achievement which was looked on with amazement."**

See the letter from Jefferson to John Jay, of April 30,

1785, cited in Roe, "English and American Tool Builders", p.

129.

** *Roe, "English and American Tool Builders", p. 133.*

While Whitney worked out his plans at Whitneyville, Simeon North, another Connecticut mechanic and a gunmaker by trade, adopted the same system. North's first shop was at Berlin. He afterwards moved to Middletown. Like Whitney, he used methods far in advance of the time. Both Whitney and North helped to establish the United States Arsenals at Springfield, Massachusetts, and at Harper's Ferry, Virginia, in which their methods were adopted. Both the Whitney and North plants survived their founders. Just before the Mexican War the Whitney plant began to use steel for gun barrels, and Jefferson Davis, Colonel of the Mississippi Rifles, declared that the new guns were "the best rifles which had ever been issued to any regiment in the world." Later, when Davis became Secretary of War, he issued to the regular army the same weapon.

The perfection of Whitney's tools and machines made it possible to employ workmen of little skill or experience. "Indeed so easy did Mr. Whitney find it to instruct new and inexperienced workmen, that he uniformly preferred to do so, rather than to combat the prejudices of those who had learned the business under a different system."* This reliance upon the machine for precision and speed has been a distinguishing mark of American manufacture. A man or a woman of little actual mechanical skill may make an excellent machine tender, learning to perform a few simple motions with great rapidity.

Denison Olmstead, "Memoir", cited by Roe, p. 159.

Whitney married in 1817 Miss Henrietta Edwards, daughter of Judge Pierpont Edwards, of New Haven, and granddaughter of Jonathan Edwards. His business prospered, and his high character, agreeable manners, and sound judgment won. for him the highest regard of all who knew him; and he had a wide circle of friends. It is said that he was on intimate terms with every President of the United States from George Washington to John Quincy Adams. But his health had been impaired by hardships endured in the South, in the long struggle over the cotton gin, and he died in 1825, at the age of fifty-nine. The business which he founded remained in his family for ninety years. It was carried on after his death by two of his nephews and then by his son, until 1888, when it was sold to the Winchester Repeating Arms Company of New Haven.

Here then, in these early New England gunshops, was born the American system of interchangeable manufacture. Its growth depended upon the machine tool, that is, the machine for making machines. Machine tools, of course, did not originate in America. English mechanics were making machines for cutting metal at least a generation before Whitney. One of the earliest of these English pioneers was John Wilkinson, inventor and maker of the boring machine which enabled Boulton and Watt in 1776 to bring their steam engine to the point of practicability. Without this machine Watt found it impossible to bore his cylinders with the necessary degree of accuracy.* From this one fact, that the success of the steam engine depended upon the invention of a new tool, we may judge of what a great part the inventors of machine tools, of whom thousands are unnamed and unknown, have played in the industrial world.

Roe, "English and American Tool Builders", p. 1 et seq.

So it was in the shops of the New England gunmakers that machine tools were first made of such variety and adaptability that they could be applied generally to other branches of manufacturing; and so it was that the system of interchangeable manufacture arose as a distinctively American development. We have already seen how England's policy of keeping at home the secrets of her machinery led to the independent development of the spindles and looms of New England. The same policy affected the tool industry in America in the same way and bred in the new country a race of original and resourceful mechanics.

One of these pioneers was Thomas Blanchard, born in 1788 on a farm in Worcester County, Massachusetts, the home also of Eli Whitney and Elias Howe. Tom began his mechanical career at the age of thirteen by inventing a device to pare apples. At the age of eighteen he went to work in his brother's shop, where tacks were made by hand, and one day took to his brother a mechanical device for counting the tacks to go into a single packet. The invention was adopted and was found to save the labor of one workman. Tom's next achievement was a machine to make tacks, on which he spent six years and the rights of which he sold for five thousand dollars. It was worth far more, for it revolutionized the tack industry, but such a sum was to young Blanchard a great fortune.

The tack-making machine gave Blanchard a reputation, and he was presently sought out by a gun manufacturer, to see whether he could improve the lathe for turning the barrels of the guns. Blanchard could; and did. His next problem was to invent a lathe for turning the irregular wooden stocks. Here he also succeeded and produced a lathe that would copy precisely and rapidly any pattern. It is from this invention that the name of Blanchard is best known. The original machine is preserved in the United States Armory at Springfield, to which Blanchard was attached for many years, and where scores of the descendants of his copying lathe may be seen in action today.

Turning gunstocks was, of course, only one of the many uses of Blanchard's copying lathe. Its chief use, in fact, was in the production of wooden lasts for the shoemakers of New England, but it was applied to many branches of wood manufacture, and later on the same principle was applied to the shaping of metal.

Blanchard was a man of many ideas. He built a steam vehicle for ordinary roads and was an early advocate of railroads; he built steamboats to ply upon the Connecticut and incidentally produced in connection with these his most profitable invention, a machine to bend ship's timbers without splintering them. The later years of his life were spent in Boston, and he often served as a patent expert in the courts, where his wide knowledge, hard common sense, incisive speech, and homely wit made him a welcome witness.

We now glance at another New England inventor, Samuel Colt, the man who carried Whitney's conceptions to transcendent heights, the most dashing and adventurous of all the pioneers of the machine shop in America. If "the American frontier was Elizabethan in quality," there was surely a touch of the Elizabethan spirit on the man whose invention so greatly affected the character of that frontier. Samuel Colt was born at Hartford in 1814 and died there in 1862 at the age of forty-eight, leaving behind him a famous name and a colossal industry of his own creation. His father was a small manufacturer of silk and woolens at Hartford, and the boy entered the factory at a very early age. At school in Amherst a little later, he fell under the displeasure of his teachers. At thirteen he took to sea, as a boy before the mast, on the East India voyage to Calcutta. It was on this voyage that he conceived the idea of the revolver and whittled out a wooden model. On his return he went into his father's works and gained a superficial knowledge of chemistry from the manager of the bleaching and dyeing department. Then he took to the road for three years and traveled from Quebec to New Orleans lecturing on chemistry under the name of "Dr. Coult." The main feature of his lecture was the administration of nitrous oxide gas to volunteers from the audience, whose antics and the amusing showman's patter made the entertainment very popular.

Colt's ambition, however, soared beyond the occupation of itinerant showman, and he never forgot his revolver. As soon as he had money enough, he made models of the new arm and took out his patents; and, having enlisted the interest of capital, he set up the Patent Arms Company at Paterson, New Jersey, to manufacture the revolver. He did not succeed in having the revolver adopted by the Government, for the army officers for a long time objected to the percussion cap (an invention, by the way, then some twenty years old, which was just coming into use and without which Colt's revolver would not have been practicable) and thought that the new weapon might fail in an emergency. Colt found a market in Texas and among the frontiersmen who were fighting the Seminole War in Florida, but the sales were insufficient, and in 1842 the company was obliged to confess insolvency and close down the plant. Colt bought from the company the patent of the revolver, which was supposed to be worthless.

Nothing more happened until after the outbreak of the Mexican War in 1846. Then came a loud call from General Zachary Taylor for a supply of Colt's revolvers. Colt had none. He had sold the last one to a Texas ranger. He had not even a model. Yet he took an order from the Government for a thousand and proceeded to construct a model. For the manufacture of the revolvers he arranged with the Whitney plant at Whitneyville. There he saw and scrutinized every detail of the factory system that Eli Whitney had established forty years earlier. He resolved to have a plant of his own on the same system and one that would far surpass Whitney's. Next year (1848) he rented premises in Hartford. His business prospered and increased. At last the Government demanded his revolvers. Within five years he had procured a site of two hundred and fifty acres fronting the Connecticut River at Hartford, and had there begun the erection of the greatest arms factory in the world.

Colt was a captain of captains. The ablest mechanic and industrial organizer in New England at that time was Elisha K. Root. Colt went after him, outbidding every other bidder for his services, and brought him to Hartford to supervise the erection of the new factory and set up its machinery. Root was a great superintendent, and the phenomenal success of the Colt factory was due in a marked degree to him. He became president of the company after Colt's death in 1862, and under him were trained a large number of mechanics and inventors of new machine tools, who afterwards became celebrated leaders and officers in the industrial armies of the country.

The spectacular rise of the Colt factory at Hartford drew the attention of the British Government, and in 1854 Colt was invited to appear in London before a Parliamentary Committee on Small Arms. He lectured the members of the committee as if they had been school boys, telling them that the regular British gun was so bad that he would be ashamed to have it come from his shop. Speaking of a plant which he had opened in London the year before he criticized the supposedly skilled British mechanic, saying: "I began here by employing the highest-priced men that I could find to do difficult things, but I had to remove the whole of these high-priced men. Then I tried the cheapest I could find, and the more ignorant a man was, the more brains he had for my purpose; and the result was this: I had men now in my employ that I started with at two shillings a day, and in one short year I can not spare them at eight shillings a day."* Colt's audacity, however, did not offend the members of the committee and they decided to visit his American factory at Hartford. They did; and were so impressed that the British Government purchased in America a full set of machines for the manufacture of arms in the Royal Small Arms factory at Enfield, England, and took across the sea American workmen and foremen to set up and run these machines. A demand sprang up in Europe for Blanchard copying lathes and a hundred other American tools, and from this time on the manufacture of tools and appliances for other manufacturers, both at home and abroad, became an increasingly important industry of New England.

Henry Barnard, "Armsmear", p. 371.

The system which the gunmakers worked out and developed to meet their own requirements was capable of indefinite expansion. It was easily adapted to other kinds of manufacture. So it was that as new inventions came in the manufacturers of these found many of the needed tools ready for them, and any special modifications could be quickly made. A manufacturer, of machine tools will produce on demand a device to perform any operation, however difficult or intricate. Some of the machines are so versatile that specially designed sets of cutting edges will adapt them to almost any work.

Standardization, due to the machine tool, is one of the chief glories of American manufacturing. Accurate watches and clocks, bicycles and motor cars, innumerable devices to save labor in the home, the office, the shop, or on the farm, are within the reach of all, because the machine tool, tended by labor comparatively unskilled, does the greater part of the work of production. In the crisis of the World War, American manufacturers, turning from the arts of peace, promptly adapted their plants to the manufacture of the most complicated engines of destruction, which were produced in Europe only by skilled machinists of the highest class.

CHAPTER IX
THE FATHERS OF ELECTRICITY

It may startle some reader to be told that the foundations of modern electrical science were definitely established in the Elizabethan Age. The England of Elizabeth, of Shakespeare, of Drake and the sea-dogs, is seldom thought of as the cradle of the science of electricity. Nevertheless, it was; just as surely as it was the birthplace of the Shakespearian drama, of the Authorized Version of the Bible, or of that maritime adventure and colonial enterprise which finally grew and blossomed into the United States of America.

The accredited father of the science of electricity and magnetism is William Gilbert, who was a physician and man of learning at the court of Elizabeth. Prior to him, all that was known of these phenomena was what the ancients knew, that the lodestone possessed magnetic properties and that amber and jet, when rubbed, would attract bits of paper or other substances of small specific gravity. Gilbert's great treatise "On the Magnet", printed in Latin in 1600, containing the fruits of his researches and experiments for many years, indeed provided the basis for a new science.

On foundations well and truly laid by Gilbert several Europeans, like Otto von Guericke of Germany, Du Fay of France, and Stephen Gray of England, worked before Benjamin Franklin and added to the structure of electrical knowledge. The Leyden jar, in which the mysterious force could be stored, was invented in Holland in 1745 and in Germany almost simultaneously.

Franklin's important discoveries are outlined in the first chapter of this book. He found out, as we have seen, that electricity and lightning are one and the same, and in the lightning rod he made the first practical application of electricity. Afterwards Cavendish of England, Coulomb of France, Galvani of Italy, all brought new bricks to the pile. Following them came a group of master builders, among whom may be mentioned: Volta of Italy, Oersted of Denmark, Ampere of France, Ohm of Germany, Faraday of England, and Joseph Henry of America.

Among these men, who were, it should be noted, theoretical investigators, rather than practical inventors like Morse, or Bell, or Edison, the American Joseph Henry ranks high. Henry was born at Albany in 1799 and was educated at the Albany Academy. Intending to practice medicine, he studied the natural sciences. He was poor and earned his daily bread by private tutoring. He was an industrious and brilliant student and soon gave evidence of being endowed with a powerful mind. He was appointed in 1824 an assistant engineer for the survey of a route for a State road, three hundred miles long, between the Hudson River and Lake Erie. The experience he gained in this work changed the course of his career; he decided to follow civil and mechanical engineering instead of medicine. Then in 1826 he became teacher of mathematics and natural philosophy in the Albany Academy.

It was in the Albany Academy that he began that wide series of experiments and investigations which touched so many phases of the great problem of electricity. His first discovery was that a magnet could be immensely strengthened by winding it with insulated wire. He was the first to employ insulated wire wound as on a spool and was able finally to make a magnet which would lift thirty-five hundred pounds. He first showed the difference between "quantity" magnets composed of short lengths of wire connected in parallel, excited by a few large cells, and "intensity" magnets wound with a single long wire and excited by a battery composed of cells in series. This was an original discovery, greatly increasing both the immediate usefulness of the magnet and its possibilities for future experiments.

The learned men of Europe, Faraday, Sturgeon, and the rest, were quick to recognize the value of the discoveries of the young Albany schoolmaster. Sturgeon magnanimously said: "Professor Henry has been enabled to produce a magnetic force which totally eclipses every other in the whole annals of magnetism; and no parallel is to be found since the miraculous suspension of the celebrated Oriental imposter in his iron coffin."*

* *Philosophical Magazine, vol. XI, p. 199 (March, 1832).*

Henry also discovered the phenomena of self induction and mutual induction. A current sent through a wire in the second story of the building induced currents through a similar wire in the cellar two floors below. In this discovery Henry anticipated Faraday though his results as to mutual induction were not published until he had heard rumors of Faraday's discovery, which he thought to be something different.

The attempt to send signals by electricity had been made many times before Henry became interested in the problem. On the invention of Sturgeon's magnet there had been hopes in England of a successful solution, but in the experiments that followed the current became so weak after a few hundred feet that the idea was pronounced impracticable. Henry strung a mile of fine wire in the Academy, placed an "intensity" battery at one end, and made the armature strike a bell at the other. Thus he discovered the essential principle of the electric telegraph. This discovery was made in 1831, the year before the idea of a working electric telegraph flashed on the mind of Morse. There was no occasion for the controversy which took place later as to who invented the telegraph. That was Morse's achievement, but the discovery of the great fact, which startled Morse into activity, was Henry's achievement. In Henry's own words: "This was the first discovery of the fact that a galvanic current could be transmitted to a great distance with so little a diminution of force as to produce mechanical effects, and of the means by which the transmission could be accomplished. I saw that the electric telegraph was now practicable." He says further, however: "I had not in mind any particular form of telegraph, but referred only to the general fact that it was now demonstrated that a galvanic current could be transmitted to great distances, with sufficient power to produce mechanical effects adequate to the desired object."*

*Deposition of Joseph Henry, September 7, 1849, printed in

Morse, "The Electra-Magnetic Telegraph", p. 91.*

Henry next turned to the possibility of a magnetic engine for the production of power and succeeded in making a reciprocating-bar motor, on which he installed the first automatic pole changer, or commutator, ever used with an electric battery. He did not succeed in producing direct rotary motion. His bar oscillated like the walking beam of a steamboat.

Henry was appointed in 1839. Professor of Natural Philosophy in the College of New Jersey, better known today as Princeton University. There he repeated his old experiments on a larger scale, confirmed Steinheil's experiment of using the earth as return conductor, showed how a feeble current would be strengthened, and how a small magnet could be used as a circuit maker and breaker. Here were the principles of the telegraph relay and the dynamo.

Why, then, if the work of Henry was so important, is his name almost forgotten, except by men of science, and not given to any one of the practical applications of electricity? The answer is plain. Henry was an investigator, not an inventor. He states his position very clearly: "I never myself attempted to reduce the principles to practice, or to apply any of my discoveries to processes in the arts. My whole attention exclusive of my duties to the College, was devoted to original scientific investigations, and I left to others what I considered in a scientific view of subordinate importance—the application of my discoveries to useful purposes in the arts. Besides this I partook of the feeling common to men of science, which disinclines them to secure to themselves the advantages of their discoveries by a patent."

Then, too, his talents were soon turned to a wider field. The bequest of James Smithson, that farsighted Englishman, who left his fortune to the United States to found "the Smithsonian Institution, for the increase and diffusion of knowledge among men," was responsible for the diffusion of Henry's activities. The Smithsonian Institution was founded at Washington in 1846, and Henry was fittingly chosen its Secretary, that is, its chief executive officer. And from that time until his death in 1878, over thirty years, he devoted himself to science in general.

He studied terrestrial magnetism and building materials. He reduced meteorology to a science, collecting reports by telegraph, made the first weather map, and issued forecasts of the weather based upon definite knowledge rather than upon signs. He became a member of the Lighthouse Board in 1852 and was the head after 1871. The excellence of marine illuminants and fog signals today is largely due to his efforts. Though he was later drawn into a controversy with Morse over the credit for the invention of the telegraph, he used his influence to procure the renewal of Morse's patent. He listened with attention to Alexander Graham Bell, who had the idea that electric wires might be made to carry the human voice, and encouraged him to proceed with his experiments. "He said," Bell writes, "that he thought it was the germ of a great invention and advised me to work at it without publishing. I said that I recognized the fact that there were mechanical difficulties in the way that rendered the plan impracticable at the present time. I added that I felt that I had not the electrical knowledge necessary to overcome the difficulties. His laconic answer was, 'GET IT!' I cannot tell you how much these two words have encouraged me."

Henry had blazed the way for others to work out the principles of the electric motor, and a few experimenters attempted to follow his lead. Thomas Davenport, a blacksmith of Brandon, Vermont, built an electric car in 1835, which he was able to drive on the road, and so made himself the pioneer of the automobile in America. Twelve years later Moses G. Farmer exhibited at various places in New England an electric-driven locomotive, and in 1851 Charles Grafton Page drove an electric car, on the tracks of the Baltimore and Ohio Railroad, from Washington to Bladensburg, at the rate of nineteen miles an hour. But the cost of batteries was too great and the use of the electric motor in transportation not yet practicable.

The great principle of the dynamo, or electric generator, was discovered by Faraday and Henry but the process of its development into an agency of practical power consumed many years; and without the dynamo for the generation of power the electric motor had to stand still and there could be no practicable application of electricity to transportation, or manufacturing, or lighting. So it was that, except for the telegraph, whose story is told in another chapter, there was little more American achievement in electricity until after the Civil War.

The arc light as a practical illuminating device came in 1878. It was introduced by Charles F. Brush, a young Ohio engineer and graduate of the University of Michigan. Others before him had attacked the problem of electric lighting, but lack of suitable carbons stood in the way of their success. Brush overcame the chief difficulties and made several lamps to burn in series from one dynamo. The first Brush lights used for street illumination were erected in Cleveland, Ohio, and soon the use of arc lights became general. Other inventors improved the apparatus, but still there were drawbacks. For outdoor lighting and for large halls they served the purpose, but they could not be used in small rooms. Besides, they were in series, that is, the current passed through every lamp in turn, and an accident to one threw the whole series out of action. The whole problem of indoor lighting was to be solved by one of America's most famous inventors.

The antecedents of Thomas Alva Edison in America may be traced back to the time when Franklin was beginning his career as a printer in Philadelphia. The first American Edisons appear to have come from Holland about 1730 and settled on the Passaic River in New Jersey. Edison's grandfather, John Edison, was a Loyalist in the Revolution who found refuge in Nova Scotia and subsequently moved to Upper Canada. His son, Samuel Edison, thought he saw a moral in the old man's exile. His father had taken the King's side and had lost his home; Samuel would make no such error. So, when the Canadian Rebellion of 1837 broke out, Samuel Edison, aged thirty-three, arrayed himself on the side of the insurgents. This time, however, the insurgents lost, and Samuel was obliged to flee to the United States, just as his father had fled to Canada. He finally settled at Milan, Ohio, and there, in 1847, in a little brick house, which is still standing, Thomas Alva Edison was born.

When the boy was seven the family moved to Port Huron, Michigan. The fact that he attended school only three months and soon became self-supporting was not due to poverty. His mother, an educated woman of Scotch extraction, taught him at home after the schoolmaster reported that he was "addled." His desire for money to spend on chemicals for a laboratory which he had fitted up in the cellar led to his first venture in business. "By a great amount of persistence," he says, "I got permission to go on the local train as newsboy. The local train from Port Huron to Detroit, a distance of sixty-three miles, left at 7 A.M. and arrived again at 9.30 P.M. After being on the train for several months I started two stores in Port Huron—one for periodicals, and the other for vegetables, butter, and berries in the season. They were attended by two boys who shared in the profits." Moreover, young Edison bought produce from the farmers' wives along the line which he sold at a profit. He had several newsboys working for him on other trains; he spent hours in the Public Library in Detroit; he fitted up a laboratory in an unused compartment of one of the coaches, and then bought a small printing press which he installed in the car and began to issue a newspaper which he printed on the train. All before he was fifteen years old.

But one day Edison's career as a traveling newsboy came to a sudden end. He was at work in his moving laboratory when a lurch of the train jarred a stick of burning phosphorus to the floor and set the car on fire. The irate conductor ejected him at the next station, giving him a violent box on the ear, which permanently injured his hearing, and dumped his chemicals and printing apparatus on the platform.

Having lost his position, young Edison soon began to dabble in telegraphy, in which he had already become interested, "probably," as he says, "from visiting telegraph offices with a chum who had tastes similar to mine." He and this chum strung a line between their houses and learned the rudiments of writing by wire. Then a station master on the railroad, whose child Edison had saved from danger, took Edison under his wing and taught him the mysteries of railway telegraphy. The boy of sixteen held positions with small stations near home for a few months and then began a period of five years of apparently purposeless wandering as a tramp telegrapher. Toledo, Cincinnati, Indianapolis, Memphis, Louisville, Detroit, were some of the cities in which he worked, studied, experimented, and played practical jokes on his associates. He was eager to learn something of the principles of electricity but found few from whom he could learn.

Edison arrived in Boston in 1868, practically penniless, and applied for a position as night operator. "The manager asked me when I was ready to go to work. 'Now,' I replied." In Boston he found men who knew something of electricity, and, as he worked at night and cut short his sleeping hours, he found time for study. He bought and studied Faraday's works. Presently came the first of his multitudinous inventions, an automatic vote recorder, for which he received a patent in 1868. This necessitated a trip to Washington, which he made on borrowed money, but he was unable to arouse any interest in the device. "After the vote recorder," he says, "I invented a stock ticker, and started a ticker service in Boston; had thirty or forty subscribers and operated from a room over the Gold Exchange." This machine Edison attempted to sell in New York, but he returned to Boston without having succeeded. He then invented a duplex telegraph by which two messages might be sent simultaneously, but at a test the machine failed because of the stupidity of the assistant.

Penniless and in debt, Edison arrived again in New York in 1869. But now fortune favored him. The Gold Indicator Company was a concern furnishing to its subscribers by telegraph the Stock Exchange prices of gold. The company's instrument was out of order. By a lucky chance Edison was on the spot to repair it, which he did successfully, and this led to his appointment as superintendent at a salary of three hundred dollars a month. When a change in the ownership of the company threw him out of the position he formed, with Franklin L. Pope, the partnership of Pope, Edison, and Company, the first firm of electrical engineers in the United States.

Not long afterwards Edison brought out the invention which set him on the high road to great achievement. This was the improved stock ticker, for which the Gold and Stock Telegraph Company paid him forty thousand dollars. It was much more than he had expected. "I had made up my mind," he says, "that, taking into consideration the time and killing pace I was working at, I should be entitled to $5000, but could get along with $3000." The money, of course, was paid by check. Edison had never received a check before and he had to be told how to cash it.

Edison immediately set up a shop in Newark and threw himself into many and various activities. He remade the prevailing system of automatic telegraphy and introduced it into England. He experimented with submarine cables and worked out a system of quadruplex telegraphy by which one wire was made to do the work of four. These two inventions were bought by Jay Gould for his Atlantic and Pacific Telegraph Company. Gould paid for the quadruplex system thirty thousand dollars, but for the automatic telegraph he paid nothing. Gould presently acquired control of the Western Union; and, having thus removed competition from his path, "he then," says Edison, "repudiated his contract with the automatic telegraph people and they never received a cent for their wires or patents, and I lost three years of very hard labor. But I never had any grudge against him because he was so able in his line, and as long as my part was successful the money with me was a secondary consideration. When Gould got the Western Union I knew no further progress in telegraphy was possible, and I went into other lines."*

* Quoted in Dyer and Martin. "Edison", vol. 1, p. 164.

In fact, however, the need of money forced Edison later on to resume his work for the Western Union Telegraph Company, both in telegraphy and telephony. His connection with the telephone is told in another volume of this series.* He invented a carbon transmitter and sold it to the Western Union for one hundred thousand dollars, payable in seventeen annual installments of six thousand dollars. He made a similar agreement for the same sum offered him for the patent of the electro-motograph. He did not realize that these installments were only simple interest upon the sums due him. These agreements are typical of Edison's commercial sense in the early years of his career as an inventor. He worked only upon inventions for which there was a possible commercial demand and sold them for a trifle to get the money to meet the pay rolls of his different shops. Later the inventor learned wisdom and associated with himself keen business men to their common profit.

* Hendrick, "The Age of Big Business".

Edison set up his laboratories and factories at Menlo Park, New Jersey, in 1876, and it was there that he invented the phonograph, for which he received the first patent in 1878. It was there, too, that he began that wonderful series of experiments which gave to the world the incandescent lamp. He had noticed the growing importance of open arc lighting, but was convinced that his mission was to produce an electric lamp for use within doors. Forsaking for the moment his newborn phonograph, Edison applied himself in earnest to the problem of the lamp. His first search was for a durable filament which would burn in a vacuum. A series of experiments with platinum wire and with various refractory metals led to no satisfactory results. Many other substances were tried, even human hair. Edison concluded that carbon of some sort was the solution rather than a metal. Almost coincidently, Swan, an Englishman, who had also been wrestling with this problem, came to the same conclusion. Finally, one day in October, 1879, after fourteen months of hard work and the expenditure of forty thousand dollars, a carbonized cotton thread sealed in one of Edison's globes lasted forty hours. "If it will burn forty hours now," said Edison, "I know I can make it burn a hundred." And so he did. A better filament was needed. Edison found it in carbonized strips of bamboo.

Edison developed his own type of dynamo, the largest ever made up to that time, and, along with the Edison incandescent lamps, it was one of the wonders of the Paris Electrical Exposition of 1881. The installation in Europe and America of plants for service followed. Edison's first great central station, supplying power for three thousand lamps, was erected at

Holborn Viaduct, London, in 1882, and in September of that year the Pearl Street Station in New York City, the first central station in America, was put into operation.

The incandescent lamp and the central power station, considered together, may be regarded as one of the most fruitful conceptions in the history of applied electricity. It comprised a complete generating, distributing, and utilizing system, from the dynamo to the very lamp at the fixture, ready for use. It even included a meter to determine the current actually consumed. The success of the system was complete, and as fast as lamps and generators could be produced they were installed to give a service at once recognized as superior to any other form of lighting. By 1885 the Edison lighting system was commercially developed in all its essentials, though still subject to many improvements and capable of great enlargement, and soon Edison sold out his interests in it and turned his great mind to other inventions.

The inventive ingenuity of others brought in time better and more economical incandescent lamps. From the filaments of bamboo fiber the next step was to filaments of cellulose in the form of cotton, duly prepared and carbonized. Later (1905) came the metalized carbon filament and finally the employment of tantalum or tungsten. The tungsten lamps first made were very delicate, and it was not until W. D. Coolidge, in the research laboratories of the General Electric Company at Schenectady, invented a process for producing ductile tungsten that they became available for general use.

The dynamo and the central power station brought the electric motor into action. The dynamo and the motor do precisely opposite things. The dynamo converts mechanical energy into electric energy. The motor transforms electric energy into mechanical energy. But the two work in partnership and without the dynamo to manufacture the power the motor could not thrive. Moreover, the central station was needed to distribute the power for transportation as well as for lighting.

The first motors to use Edison station current were designed by Frank J. Sprague, a graduate of the Naval Academy, who had worked with Edison, as have many of the foremost electrical engineers of America and Europe. These small motors possessed several advantages over the big steam engine. They ran smoothly and noiselessly on account of the absence of reciprocating parts. They consumed current only when in use. They could be installed and connected with a minimum of trouble and expense. They emitted neither smell nor smoke. Edison built an experimental electric railway line at Menlo Park in 1880 and proved its practicability. Meanwhile, however, as he worked on his motors and dynamos, he was anticipated by others in some of his inventions. It would not be fair to say that Edison and Sprague alone developed the electric railway, for there were several others who made important contributions. Stephen D. Field of Stockbridge, Massachusetts, had a patent which the Edison interests found it necessary to acquire; C. J. Van Depoele and Leo Daft made important contributions to the trolley system. In Cleveland in 1884 an electric railway on a small scale was opened to the public. But Sprague's first electric railway, built at Richmond, Virginia, in 1887, as a complete system, is generally hailed as the true pioneer of electric transportation in the United States. Thereafter the electric railway spread quickly over the land, obliterating the old horsecars and greatly enlarging the circumference of the city. Moreover, on the steam roads, at all the great terminals, and wherever there were tunnels to be passed through, the old giant steam engine in time yielded place to the electric motor.

The application of the electric motor to the "vertical railway," or elevator, made possible the steel skyscraper. The elevator, of course, is an old device. It was improved and developed in America by Elisha Graves Otis, an inventor who lived and died before the Civil War and whose sons afterward erected a great business on foundations laid by him. The first Otis elevators were moved by steam or hydraulic power. They were slow, noisy, and difficult of control. After the electric motor came in; the elevator soon changed its character and adapted itself to the imperative demands of the towering, skeleton-framed buildings which were rising in every city.

Edison, already famous as "the Wizard of Menlo Park," established his factories and laboratories at West Orange, New Jersey, in 1887, whence he has since sent forth a constant stream of inventions, some new and startling, others improvements on old devices. The achievements of several other inventors in the electrical field have been only less noteworthy than his. The new profession of electrical engineering called to its service great numbers of able men. Manufacturers of electrical machinery established research departments and employed inventors. The times had indeed changed since the day when Morse, as a student at Yale College, chose art instead of electricity as his calling, because electricity afforded him no means of livelihood.

From Edison's plant in 1903 came a new type of the storage battery, which he afterwards improved. The storage battery, as every one knows, is used in the propulsion of electric vehicles and boats, in the operation of block-signals, in the lighting of trains, and in the ignition and starting of gasoline engines. As an adjunct of the gas-driven automobile, it renders the starting of the engine independent of muscle and so makes possible the general use of the automobile by women as well as men.

The dynamo brought into service not only light and power but heat; and the electric furnace in turn gave rise to several great metallurgical and chemical industries. Elihu Thomson's process of welding by means of the arc furnace found wide and varied applications. The commercial production of aluminum is due to the electric furnace and dates from 1886. It was in that year that H. Y. Castner of New York and C. M. Hall of Pittsburgh both invented the methods of manufacture which gave to the world the new metal, malleable and ductile, exceedingly light, and capable of a thousand uses. Carborundum is another product of the electric furnace. It was the invention of Edward B. Acheson, a graduate of the Edison laboratories. Acheson, in 1891, was trying to make artificial diamonds and produced instead the more useful carborundum, as well as the Acheson graphite, which at once found its place in industry. Another valuable product of the electric furnace was the calcium carbide first produced in 1892 by Thomas L. Wilson of Spray, North Carolina. This calcium carbide is the basis of acetylene gas, a powerful illuminant, and it is widely used in metallurgy, for welding and other purposes.

At the same time with these developments the value of the alternating current came to be recognized. The transformer, an instrument developed on foundations laid by Henry and Faraday, made it possible to transmit electrical energy over great distances with little loss of power. Alternating currents were transformed by means of this instrument at the source, and were again converted at the point of use to a lower and convenient potential for local distribution and consumption. The first extensive use of the alternating current was in arc lighting, where the higher potentials could be employed on series lamps. Perhaps the chief American inventor in the domain of the alternating current is Elihu Thomson, who began his useful career as Professor of Chemistry and Mechanics in the Central High School of Philadelphia. Another great protagonist of the alternating current was George Westinghouse, who was quite as much an improver and inventor as a manufacturer of machinery. Two other inventors, at least, should not be forgotten in this connection: Nicola Tesla and Charles S. Bradley. Both of them had worked for Edison.

The turbine (from the Latin turbo, meaning a whirlwind) is the name of the motor which drives the great dynamos for the generation of electric energy. It may be either a steam turbine or a water turbine. The steam turbine of Curtis or Parsons is today the prevailing engine. But the development of hydro-electric power has already gone far. It is estimated that the electric energy produced in the United States by the utilization of water powers every year equals the power product of forty million tons of coal, or about one-tenth of the coal which is consumed in the production of steam. Yet hydro-electricity is said to be only in its beginnings, for not more than a tenth of the readily available water power of the country is actually in use.

The first commercial hydro-station for the transmission of power in America was established in 1891 at Telluride, Colorado. It was practically duplicated in the following year at Brodie, Colorado. The motors and generators for these stations came from the Westinghouse plant in Pittsburgh, and Westinghouse also supplied the turbo-generators which inaugurated, in 1895, the delivery of power from Niagara Falls.

CHAPTER X
THE CONQUEST OF THE AIR

The most popular man in Europe in the year 1783 was still the United States Minister to France. The figure of plain Benjamin Franklin, his broad head, with the calm, shrewd eyes peering through the bifocals of his own invention, invested with a halo of great learning and fame, entirely captivated the people's imagination.

As one of the American Commissioners busy with the extraordinary problems of the Peace, Franklin might have been supposed too occupied for excursions into the paths of science and philosophy. But the spaciousness and orderly furnishing of his mind provided that no pursuit of knowledge should be a digression for him. So we find him, naturally, leaving his desk on several days of that summer and autumn and posting off to watch the trials of a new invention; nothing less indeed than a ship to ride the air. He found time also to describe the new invention in letters to his friends in different parts of the world.

On the 21st of November Franklin set out for the gardens of the King's hunting lodge in the Bois de Boulogne, on the outskirts of Paris, with a quickened interest, a thrill of excitement, which made him yearn to be young again with another long life to live that he might see what should be after him on the earth. What bold things men would attempt! Today two daring Frenchmen, Pilatre de Rozier of the Royal Academy and his friend the Marquis d'Arlandes, would ascend in a balloon freed from the earth—the first men in history to adventure thus upon the wind. The crowds gathered to witness the event opened a lane for Franklin to pass through.

At six minutes to two the aeronauts entered the car of their balloon; and, at a height of two hundred and seventy feet, doffed their hats and saluted the applauding spectators. Then the wind carried them away toward Paris. Over Passy, about half a mile from the starting point, the balloon began to descend, and the River Seine seemed rising to engulf them; but when they fed the fire under their sack of hot air with chopped straw they rose to the elevation of five hundred feet. Safe across the river they dampened the fire with a sponge and made a gentle descent beyond the old ramparts of Paris.

At five o'clock that afternoon, at the King's Chateau in the Bois de Boulogne, the members of the Royal Academy signed a memorial of the event. One of the spectators accosted Franklin.

"What does Dr. Franklin conceive to be the use of this new invention?"

"What is the use of a new-born child?" was the retort.

A new-born child, a new-born republic, a new invention: alike dim beginnings of development which none could foretell. The year that saw the world acknowledge a new nation, freed of its ancient political bonds, saw also the first successful attempt to break the supposed bonds that held men down to the ground. Though the invention of the balloon was only five months old, there were already two types on exhibition: the original Montgolfier, or fireballoon, inflated with hot air, and a modification by Charles, inflated with hydrogen gas. The mass of the French people did not regard these balloons with Franklin's serenity. Some weeks earlier the danger of attack had necessitated a balloon's removal from the place of its first moorings to the Champ de Mars at dead of night. Preceded by flaming torches, with soldiers marching on either side and guards in front and rear, the great ball was borne through the darkened streets. The midnight cabby along the route stopped his nag, or tumbled from sleep on his box, to kneel on the pavement and cross himself against the evil that might be in that strange monster. The fear of the people was so great that the Government saw fit to issue a proclamation, explaining the invention. Any one seeing such a globe, like the moon in an eclipse, so read the proclamation, should be aware that it is only a bag made of taffeta or light canvas covered with paper and "cannot possibly cause any harm and which will some day prove serviceable to the wants of society."

Franklin wrote a description of the Montgolfier balloon to Sir Joseph Banks, President of the Royal Society of London:

"Its bottom was open and in the middle of the opening was fixed a kind of basket grate, in which faggots and sheaves of straw were burnt. The air, rarefied in passing through this flame, rose in the balloon, swelled out its sides, and filled it. The persons, who were placed in the gallery made of wicker and attached to the outside near the bottom, had each of them a port through which they could pass sheaves of straw into the grate to keep up the flame and thereby keep the balloon full.... One of these courageous philosophers, the Marquis d'Arlandes, did me the honor to call upon me in the evening after the experiment, with Mr. Montgolfier, the very ingenious inventor. I was happy to see him safe. He informed me that they lit gently, without the least shock, and the balloon was very little damaged."

Franklin writes that the competition between Montgolfier and Charles has already resulted in progress in the construction and management of the balloon. He sees it as a discovery of great importance, one that "may possibly give a new turn to human affairs. Convincing sovereigns of the folly of war may perhaps be one effect of it, since it will be impracticable for the most potent of them to guard his dominions." The prophecy may yet be fulfilled. Franklin remarks that a short while ago the idea of "witches riding through the air upon a broomstick and that of philosophers upon a bag of smoke would have appeared equally impossible and ridiculous." Yet in the space of a few months he has seen the philosopher on his smoke bag, if not the witch on her broom. He wishes that one of these very ingenious inventors would immediately devise means of direction for the balloon, a rudder to steer it; because the malady from which he is suffering is always increased by a jolting drive in a fourwheeler and he would gladly avail himself of an easier way of locomotion.

The vision of man on the wing did not, of course, begin with the invention of the balloon. Perhaps the dream of flying man came first to some primitive poet of the Stone Age, as he watched, fearfully, the gyrations of the winged creatures of the air; even as in a later age it came to Langley and Maxim, who studied the wing motions of birds and insects, not in fear but in the light and confidence of advancing science.

Crudely outlined by some ancient Egyptian sculptor, a winged human figure broods upon the tomb of Rameses III. In the Hebrew parable of Genesis winged cherubim guarded the gates of Paradise against the man and woman who had stifled aspiration with sin. Fairies, witches, and magicians ride the wind in the legends and folklore of all peoples. The Greeks had gods and goddesses many; and one of these Greek art represents as moving earthward on great spreading pinions. Victory came by the air. When Demetrius, King of Macedonia, set up the Winged Victory of Samothrace to commemorate the naval triumph of the Greeks over the ships of Egypt, Greek art poetically foreshadowed the relation of the air service to the fleet in our own day.

Man has always dreamed of flight; but when did men first actually fly? We smile at the story of Daedalus, the Greek architect, and his son, Icarus, who made themselves wings and flew from the realm of their foes; and the tale of Simon, the magician, who pestered the early Christian Church by exhibitions of flight into the air amid smoke and flame in mockery of the ascension. But do the many tales of sorcerers in the Middle Ages, who rose from the ground with their cloaks apparently filled with wind, to awe the rabble, suggest that they had deduced the principle of the aerostat from watching the action of smoke as did the Montgolfiers hundreds of years later? At all events one of these alleged exhibitions about the year 800 inspired the good Bishop Agobard of Lyons to write a book against superstition, in which he proved conclusively that it was impossible for human beings to rise through the air. Later, Roger Bacon and Leonardo da Vinci, each in his turn ruminated in manuscript upon the subject of flight. Bacon, the scientist, put forward a theory of thin copper globes filled with liquid fire, which would soar. Leonardo, artist, studied the wings of birds. The Jesuit Francisco Lana, in 1670, working on Bacon's theory sketched an airship made of four copper balls with a skiff attached; this machine was to soar by means of the lighter-than-air globes and to be navigated aloft by oars and sails.

But while philosophers in their libraries were designing airships on paper and propounding their theories, venturesome men, "crawling, but pestered with the thought of wings," were making pinions of various fabrics and trying them upon the wind. Four years after Lana suggested his airship with balls and oars, Besnier, a French locksmith, made a flying machine of four collapsible planes like book covers suspended on rods. With a rod over each shoulder, and moving the two front planes with his arms and the two back ones by his feet, Besnier gave exhibitions of gliding from a height to the earth. But his machine could not soar. What may be called the first patent on a flying machine was recorded in 1709 when Bartholomeo de Gusmao, a friar, appeared before the King of Portugal to announce that he had invented a flying machine and to request an order prohibiting other men from making anything of the sort. The King decreed pain of death to all infringers; and to assist the enterprising monk in improving his machine, he appointed him first professor of mathematics in the University of Coimbra with a fat stipend. Then the Inquisition stepped in. The inventor's suave reply, to the effect that to show men how to soar to Heaven was an essentially religious act, availed him nothing. He was pronounced a sorcerer, his machine was destroyed, and he was imprisoned till his death. Many other men fashioned unto themselves wings; but, though some of them might glide earthward, none could rise upon the wind.

While the principle by which the balloon, father of the dirigible, soars and floats could be deduced by men of natural powers of observation and little science from the action of clouds and smoke, the airplane, the Winged Victory of our day, waited upon two things—the scientific analysis of the anatomy of bird wings and the internal combustion engine.

These two things necessary to convert man into a rival of the albatross did not come at once and together. Not the dream of flying but the need for quantity and speed in production to take care of the wants of a modern civilization compelled the invention of the internal combustion engine.

Before it appeared in the realm of mechanics, experimenters were applying in the construction of flying models the knowledge supplied by Cayley in 1796, who made an instrument of whalebone, corks, and feathers, which by the action of two screws of quill feathers, rotating in opposite directions, would rise to the ceiling; and the full revelation of the structure and action of bird wings set forth by Pettigrew in 1867.

"The wing, both when at rest and when in motion," Pettigrew declared, "may not inaptly be compared to the blade of an ordinary screw propeller as employed in navigation. Thus the general outline of the wing corresponds closely with the outline of the propeller, and the track described by the wing in space IS TWISTED UPON ITSELF propeller fashion." Numerous attempts to apply the newly discovered principles to artificial birds failed, yet came so close to success that they fed instead of killing the hope that a solution of the problem would one day ere long be reached.

"Nature has solved it, and why not man?"

From his boyhood days Samuel Pierpont Langley, so he tells us, had asked himself that question, which he was later to answer. Langley, born in Roxbury, Massachusetts, in 1834, was another link in the chain of distinguished inventors who first saw the light of day in Puritan New England. And, like many of those other inventors, he numbered among his ancestors for generations two types of men—on the one hand, a line of skilled artisans and mechanics; on the other, the most intellectual men of their time such as clergymen and schoolmasters, one of them being Increase Mather. We see in Langley, as in some of his brother New England inventors, the later flowering of the Puritan ideal stripped of its husk of superstition and harshness—a high sense of duty and of integrity, an intense conviction that the reason for a man's life here is that he may give service, a reserved deportment which did not mask from discerning eyes the man's gentle qualities of heart and his keen love of beauty in art and Nature.

Langley first chose as his profession civil engineering and architecture and the years between 1857 and 1864 were chiefly spent in prosecuting these callings in St. Louis and Chicago. Then he abandoned them; for the bent of his mind was definitely towards scientific inquiry. In 1867 he was appointed director of the Allegheny Observatory at Pittsburgh. Here he remained until 1887, when, having made for himself a world-wide reputation as an astronomer, he became Secretary of the Smithsonian Institution at Washington.

It was about this time that he began his experiments in "aerodynamics." But the problem of flight had long been a subject of interested speculation with him. Ten years later he wrote:

"Nature has made her flying-machine in the bird, which is nearly a thousand times as heavy as the air its bulk displaces, and only those who have tried to rival it know how inimitable her work is, for the "way of a bird in the air" remains as wonderful to us as it was to Solomon, and the sight of the bird has constantly held this wonder before men's minds, and kept the flame of hope from utter extinction, in spite of long disappointment. I well remember how, as a child, when lying in a New England pasture, h watched a hawk soaring far up in the blue, and sailing for a long time without any motion of its wings, as though it needed no work to sustain it, but was kept up there by some miracle. But, however sustained, I saw it sweep in a few seconds of its leisurely flight, over a distance that to me was encumbered with every sort of obstacle, which did not exist for it.... How wonderfully easy, too, was its flight! There was not a flutter of its pinions as it swept over the field, in a motion which seemed as effortless as that of its shadow. After many years and in mature life, I was brought to think of these things again, and to ask myself whether the problem of artificial flight was as hopeless and as absurd as it was then thought to be"... In three or four years Langley made nearly forty models. "The primary difficulty lay in making the model light enough and sufficiently strong to support its power," he says. "This difficulty continued to be fundamental through every later form; but, beside this, the adjustment of the center of gravity to the center of pressure of the wings, the disposition of the wings themselves, the size of the propellers, the inclination and number of the blades, and a great number of other details, presented themselves for examination."

By 1891 Langley had a model light enough to fly, but proper balancing had not been attained. He set himself anew to find the practical conditions of equilibrium and of horizontal flight. His experiments convinced him that "mechanical sustenation of heavy bodies in the air, combined with very great speeds, is not only possible, but within the reach of mechanical means we actually possess."

After many experiments with new models Langley at length fashioned a steam-driven machine which would fly horizontally. It weighed about thirty pounds; it was some sixteen feet in length, with two sets of wings, the pair in front measuring forty feet from tip to tip. On May 6, 1896, this model was launched over the Potomac River. It flew half a mile in a minute and a half. When its fuel and water gave out, it descended gently to the river's surface. In November Langley launched another model which flew for three-quarters of a mile at a speed of thirty miles an hour. These tests demonstrated the practicability of artificial flight.

The Spanish-American War found the military observation balloon doing the limited work which it had done ever since the days of Franklin.

President McKinley was keenly interested in Langley's design to build a power-driven flying machine which would have innumerable advantages over the balloon. The Government provided the funds and Langley took up the problem of a flying machine large enough to carry a man. His initial difficulty was the engine. It was plain at once that new principles of engine construction must be adopted before a motor could be designed of high power yet light enough to be borne in the slender body of an airplane. The internal combustion engine had now come into use. Langley went to Europe in 1900, seeking his motor, only to be told that what he sought was impossible.

His assistant, Charles M. Manly, meanwhile found a builder of engines in America who was willing to make the attempt. But, after two years of waiting for it, the engine proved a failure. Manly then had the several parts of it, which he deemed hopeful, transported to Washington, and there at the Smithsonian Institution he labored and experimented until he evolved a light and powerful gasoline motor. In October, 1903, the test was made, with Manly aboard of the machine. The failure which resulted was due solely to the clumsy launching apparatus. The airplane was damaged as it rushed forward before beginning to soar; and, as it rose, it turned over and plunged into the river. The loyal and enthusiastic Manly, who was fortunately a good diver and swimmer, hastily dried himself and gave out a reassuring statement to the representatives of the press and to the officers of the Board of Ordnance gathered to witness the flight.

A second failure in December convinced spectators that man was never intended to fly. The newspapers let loose such a storm of ridicule upon Langley and his machine, with charges as to the waste of public funds, that the Government refused to assist him further. Langley, at that time sixty-nine years of age, took this defeat so keenly to heart that it hastened his death, which occurred three years later. "Failure in the aerodrome itself," he wrote, "or its engines there has been none; and it is believed that it is at the moment of success, and when the engineering problems have been solved, that a lack of means has prevented a continuance of the work."

It was truly "at the moment of success" that Langley's work was stopped. On December 17, 1903, the Wright brothers made the first successful experiment in which a machine carrying a man rose by its own power, flew naturally and at even speed, and descended without damage. These brothers, Wilbur and Orville, who at last opened the long besieged lanes of the air, were born in Dayton, Ohio. Their father, a clergyman and later a bishop, spent his leisure in scientific reading and in the invention of a typewriter which, however, he never perfected. He inspired an interest in scientific principles in his boys' minds by giving them toys which would stimulate their curiosity. One of these toys was a helicopter, or Cayley's Top, which would rise and flutter awhile in the air.

After several helicopters of their own, the brothers made original models of kites, and Orville, the younger, attained an exceptional skill in flying them. Presently Orville and Wilbur were making their own bicycles and astonishing their neighbors by public appearances on a specially designed tandem. The first accounts which they read of experiments with flying machines turned their inventive genius into the new field. In particular the newspaper accounts at that time of Otto Lilienthal's exhibitions with his glider stirred their interest and set them on to search the libraries for literature on the subject of flying. As they read of the work of Langley and others they concluded that the secret of flying could not be mastered theoretically in a laboratory; it must be learned in the air. It struck these young men, trained by necessity to count pennies at their full value, as "wasteful extravagance" to mount delicate and costly machinery on wings which no one knew how to manage. They turned from the records of other inventors' models to study the one perfect model, the bird. Said Wilbur Wright, speaking before the Society of Western Engineers, at Chicago:

"The bird's wings are undoubtedly very well designed indeed, but it is not any extraordinary efficiency that strikes with astonishment, but rather the marvelous skill with which they are used. It is true that I have seen birds perform soaring feats of almost incredible nature in positions where it was not possible to measure the speed and trend of the wind, but whenever it was possible to determine by actual measurements the conditions under which the soaring was performed it was easy to account for it on the basis of the results obtained with artificial wings. The soaring problem is apparently not so much one of better wings as of better operators."*

Cited in Turner, "The Romance of Aeronautics".

When the Wrights determined to fly, two problems which had beset earlier experimenters had been partially solved. Experience had brought out certain facts regarding the wings; and invention had supplied an engine. But the laws governing the balancing and steering of the machine were unknown. The way of a man in the air had yet to be discovered.

The starting point of their theory of flight seems to have been that man was endowed with an intelligence at least equal to that of the bird; and, that with practice he could learn to balance himself in the air as naturally and instinctively as on the ground. He must and could be, like the bird, the controlling intelligence of his machine. To quote Wilbur Wright again:

"It seemed to us that the main reason why the problem had remained so long unsolved was that no one had been able to obtain any adequate practice. Lilienthal in five years of time had spent only five hours in actual gliding through the air. The wonder was not that he had done so little but that he had accomplished so much. It would not be considered at all safe for a bicycle rider to attempt to ride through a crowded city street after only five hours' practice spread out in bits of ten seconds each over a period of five years, yet Lilienthal with his brief practice was remarkably successful in meeting the fluctuations and eddies of wind gusts. We thought that if some method could be found by which it would be possible to practice by the hour instead of by the second, there would be a hope of advancing the solution of a very difficult problem."

The brothers found that winds of the velocity they desired for their experiments were common on the coast of North Carolina. They pitched their camp at Kitty Hawk in October, 1900, and made a brief and successful trial of their gliding machine. Next year, they returned with a much larger machine; and in 1902 they continued their experiments with a model still further improved from their first design. Having tested their theories and become convinced that they were definitely on the right track, they were no longer satisfied merely to glide. They set about constructing a power machine. Here a new problem met them. They had decided on two screw propellers rotating in opposite directions on the principle of wings in flight; but the proper diameter, pitch, and area of blade were not easily arrived at.

On December 17, 1903, the first Wright biplane was ready to navigate the air and made four brief successful flights. Subsequent flights in 1904 demonstrated that the problem of equilibrium had not been fully solved; but the experiments of 1905 banished this difficulty.

The responsibility which the Wrights placed upon the aviator for maintaining his equilibrium, and the tailless design of their machine, caused much headshaking among foreign flying men when Wilbur Wright appeared at the great aviation meet in France in 1908. But he won the Michelin Prize of eight hundred pounds by beating previous records for speed and for the time which any machine had remained in the air. He gave exhibitions also in Germany and Italy and instructed Italian army officers in the flying of Wright machines. At this time Orville was giving similar demonstrations in America. Transverse control, the warping device invented by the Wright brothers for the preservation of lateral balance and for artificial inclination in making turns, has been employed in a similar or modified form in most airplanes since constructed.

There was no "mine" or "thine" in the diction of the Wright brothers; only "we" and "ours." They were joint inventors; they shared their fame equally and all their honors and prizes also until the death of Wilbur in

1912. They were the first inventors to make the ancient dream of flying man a reality and to demonstrate that reality to the practical world.

When the NC flying boats of the United States navy lined up at Trepassey in May, 1919, for their Atlantic venture, and the press was full of pictures of them, how many hasty readers, eager only for news of the start, stopped to think what the initials NC stood for?

The seaplane is the chief contribution of Glenn Hammond Curtiss to aviation, and the Navy Curtiss Number Four, which made the first transatlantic flight in history, was designed by him. The spirit of cooperation, expressed in pooling ideas and fame, which the Wright brothers exemplified, is seen again in the association of Curtiss with the navy during the war. NC is a fraternity badge signifying equal honors.

Curtiss, in 1900, was—like the Wrights—the owner of a small bicycle shop. It was at Hammondsport, New York. He was an enthusiastic cyclist, and speed was a mania with him. He evolved a motor cycle with which he broke all records for speed over the ground. He started a factory and achieved a reputation for excellent motors. He designed and made the engine for the dirigible of Captain Thomas S. Baldwin; and for the first United States army dirigible in 1905.

Curtiss carried on some of his experiments in association with Alexander Graham Bell, who was trying to evolve a stable flying machine on the principle of the cellular kite. Bell and Curtiss, with three others, formed in 1907, the Aerial Experimental Association at Bell's country house in Canada, which was fruitful of results, and Curtiss scored several notable triumphs with the craft they designed. But the idea of a machine which could descend and propel itself on water possessed his mind, and in 1911 he exhibited at the aviation meet in Chicago the hydroaeroplane. An incident there set him dreaming of the life-saving systems on great waters. His hydroaeroplane had just returned to its hangar, after a series of maneuvers, when a monoplane in flight broke out of control and plunged into Lake Michigan. The Curtiss machine left its hangar on the minute, covered the intervening mile, and alighted on the water to offer aid. The presence of boats made the good offices of the hydroaeroplane unnecessary on that occasion; but the incident opened up to the mind of Curtiss new possibilities.

In the first years of the World War Curtiss built airplanes and flying boats for the Allies. The United States entered the arena and called for his services. The Navy Department called for the big flying boat; and the NC type was evolved, which, equipped with four Liberty Motors, crossed the Atlantic after the close of the war.

The World War, of course, brought about the magical development of all kinds of air craft. Necessity not only mothered invention but forced it to cover a normal half century of progress in four years. While Curtiss worked with the navy, the Dayton-Wright factory turned out the famous DH fighting planes under the supervision of Orville Wright. The second initial here stands for Havilland, as the DH was designed by Geoffrey de Havilland, a British inventor.

The year 1919 saw the first transatlantic flights. The NC4, with Lieutenant Commander Albert Cushing Read and crew, left Trepassey, Newfoundland, on the 16th of May and in twelve hours arrived at Horta, the Azores, more than a thousand miles away. All along the course the navy had strung a chain of destroyers, with signaling apparatus and searchlights to guide the aviators. On the twenty-seventh, NC4 took off from San Miguel, Azores, and in nine hours made Lisbon—Lisbon, capital of Portugal, which sent out the first bold mariners to explore the Sea of Darkness, prior to Columbus. On the thirtieth, NC4 took off for Plymouth, England, and arrived in ten hours and twenty minutes. Perhaps a phantom ship, with sails set and flags blowing, the name Mayflower on her hull, rode in Plymouth Harbor that day to greet a New England pilot.

On the 14th of June the Vickers-Vimy Rolls-Royce biplane, piloted by John Alcock and with Arthur Whitten Brown as observer-navigator, left St. John's, Newfoundland, and arrived at Clifden, Ireland, in sixteen hours twelve minutes, having made the first non-stop transatlantic flight. Hawker and Grieve meanwhile had made the same gallant attempt in a single-engined Sopwith machine; and had come down in mid-ocean, after flying fourteen and a half hours, owing to the failure of their water circulation. Their rescue by slow Danish Mary completed a fascinating tale of heroic adventure. The British dirigible R34, with Major G. H. Scott in command, left East Fortune, Scotland, on the 2d of July, and arrived at Mineola, New York, on the sixth. The R34 made the return voyage in seventy-five hours. In November, 1919, Captain Sir Ross Smith set off from England in a biplane to win a prize of ten thousand pounds offered by the Australian Commonwealth to the first Australian aviator to fly from England to Australia in thirty days. Over France, Italy, Greece, over the Holy Land, perhaps over the Garden of Eden, whence the winged cherubim drove Adam and Eve, over Persia, India, Siam, the Dutch East Indies to Port Darwin in northern Australia; and then southeastward across Australia itself to Sydney, the biplane flew without mishap. The time from Hounslow, England, to Port Darwin was twenty-seven days, twenty hours, and twenty minutes. Early in 1920 the Boer airman Captain Van Ryneveld made the flight from Cairo to the Cape.

Commercial development of the airplane and the airship commenced after the war. The first air service for United States mails was, in fact, inaugurated during the war, between New York and Washington. The transcontinental service was established soon afterwards, and a regular line between Key West and Havana. French and British companies began to operate daily between London and Paris carrying passengers and mail. Airship companies were formed in Australia, South Africa, and India. In Canada airplanes were soon being used in prospecting the Labrador timber regions, in making photographs and maps of the northern wilderness, and by the Northwest Mounted Police.

It is not for history to prophesy. "Emblem of much, and of our Age of Hope itself," Carlyle called the balloon of his time, born to mount majestically but "unguidably" only to tumble "whither Fate will." But the aircraft of our day is guidable, and our Age of Hope is not rudderless nor at the mercy of Fate.

BIBLIOGRAPHICAL NOTE

GENERAL

A clear, non-technical discussion of the basis of all industrial progress is "Power", by Charles E. Lucke (1911), which discusses the general principle of the substitution of power for the labor of men. Many of the references given in "Colonial Folkways", by C. M. Andrews ("The Chronicles of America", vol. IX), are valuable for an understanding of early industrial conditions. The general course of industry and commerce in the United States is briefly told by Carroll D. Wright in "The Industrial Evolution of the United States" (1907), by E. L. Bogart in "The Economic History of the United States" (1920), and by Katharine Coman in "The Industrial History of the United States" (1911). "A Documentary History of American Industrial Society", 10 vols. (1910-11), edited by John R. Commons, is a mine of material. See also Emerson D. Fite, "Social and Industrial Conditions in the North During the Civil War" (1910). The best account of the inventions of the nineteenth century is "The Progress of Invention in the Nineteenth Century" by Edward W. Byrn (1900). George Iles in "Leading American Inventors" (1912) tells the story of several important inventors and their work. The same author in "Flame, Electricity and the Camera" (1900) gives much valuable information.

CHAPTER I

The primary source of information on Benjamin Franklin is contained in his own writings. These were compiled and edited by Jared Sparks, "The Works of... Franklin... with Notes and a Life of the Author", 10 vols. (1836-40); and later by John Bigelow, "The Complete Works of Benjamin Franklin; including His Private as well as His Official and Scientific Correspondence, and Numerous Letters and Documents Now for the First Time Printed, with Many Others not included in Any Former Collection, also, the Unmutilated and Correct Version of His Autobiography", 10 vols. (1887-88). Consult also James Parton, "The Life and Times of Benjamin Franklin", 2 vols. (1864); S. G. Fisher, "The True Benjamin Franklin" (1899); Paul Leicester Ford, "The Many-Sided Franklin" (1899); John T. Morse, "Benjamin Franklin" (1889) in the "American Statesmen" series; and Lindsay Swift, "Benjamin Franklin" (1910) in "Beacon Biographies. On the Patent Office: Henry L. Ellsworth, A Digest of Patents Issued by the United States from 1790 to January 1, 1839" (Washington, 1840); also the regular Reports and publications of the United States Patent Office.

CHAPTER II

The first life of Eli Whitney is the "Memoir" by Denison Olmsted (1846), and a collection of Whitney's letters about the cotton gin may be found in "The American Historical Review", vol. III (1897). "Eli Whitney and His Cotton Gin," by M. F. Foster, is included in the "Transactions of the New England Cotton Manufacturers' Association", no. 67 (October, 1899). See also Dwight Goddard, "A Short Story of Eli Whitney" (1904); D. A. Tompkins, "Cotton and Cotton Oil" (1901); James A. B. Scherer, "Cotton as a World Power" (1916); E. C. Bates, "The Story of the Cotton Gin" (1899), reprinted from "The New England Magazine", May, 1890; and Eugene Clyde Brooks, "The Story of Cotton and the Development of the Cotton States" (1911).

CHAPTER III

For an account of James Watt's achievements, see J. Cleland, "Historical Account of the Steam Engine" (1825) and John W. Grant, "Watt and the Steam Age" (1917). On Fulton: R. H. Thurston, "Robert Fulton" (1891) in the "Makers of America" series; A. C. Sutcliffe, "Robert Fulton and the 'Clermont'" (1909); H. W. Dickinson, "Robert Fulton, Engineer and Artist; His Life and Works" (1913). For an account of John Stevens, see George Iles, "Leading American Inventors" (1912), and Dwight Goddard, "A Short Story of John Stevens and His Sons in Eminent Engineers" (1905). See also John Stevens, "Documents Tending to Prove the Superior Advantages of Rail-Ways and Steam-Carriages over Canal Navigation" (1819.), reprinted in "The Magazine of History with Notes and Queries", Extra Number 54 (1917). On Evans: "Oliver Evans and His Inventions," by Coleman Sellers, in "The Journal of the Franklin Institute", July, 1886, vol. CXXII.

CHAPTER IV

On the general subject of cotton manufacture and machinery, see: J. L. Bishop, "History of American Manufactures from 1608 to 1860", 3 vols. (1864-67); Samuel Batchelder, "Introduction and Early Progress of the Cotton Manufacture in the United States" (1863); James Montgomery, "A Practical Detail of the Cotton Manufacture of the United States of America" (1840); Melvin T. Copeland, "The Cotton Manufacturing Industry of the United States" (1912); and John L. Hayes, "American Textile Machinery"

(1879). Harriet H. Robinson, "Loom and Spindle" (1898), is a description of the life of girl workers in the early factories written by one of them. Charles Dickens, "American Notes", Chapter IV, is a vivid account of the life in the Lowell mills. See also Nathan Appleton, "Introduction of the Power Loom and Origin of Lowell" (1858); H. A. Miles, "Lowell, as It Was, and as It Is" (1845), and G. S. White, "Memoir of Samuel Slater" (1836). On Elias Howe, see Dwight Goddard, "A Short Story of Elias Howe in Eminent Engineers" (1905).

CHAPTER V

The story of the reaper is told in: Herbert N. Casson, "Cyrus Hall McCormick; His Life and Work" (1909), and "The Romance of the Reaper" (1908), and Merritt F. Miller, "Evolution of Reaping Machines" (1902), U. S. Experiment Stations Office, Bulletin 103. Other farm inventions are covered in: William Macdonald, "Makers of Modern Agriculture" (1913); Emile Guarini, "The Use of Electric Power in Plowing" in The "Electrical Review", vol. XLIII; A. P. Yerkes, "The Gas Tractor in Eastern Farming" (1918), U. S. Department of Agriculture, Farmer's Bulletin 1004; and Herbert N. Casson and others, "Horse, Truck and Tractor; the Coming of Cheaper Power for City and Farm" (1913).

CHAPTER VI

An account of an early "agent of communication" is given by W. F. Bailey, article on the "Pony Express" in "The Century Magazine", vol. XXXIV (1898). For the story of the telegraph and its inventors, see: S. I. Prime, "Life of Samuel F. B. Morse" (1875); S. F. B. Morse, "The Electro-Magnetic Telegraph" (1858) and "Examination of the Telegraphic Apparatus and the Process in Telegraphy" (1869); Guglielmo Marconi, "The Progress of Wireless Telegraphy" (1912) in the "Transactions of the New York Electrical Society", no. 15; and Ray Stannard Baker, "Marconi's Achievement" in McClure's Magazine, vol. XVIII (1902). On the telephone, see Herbert N. Casson, "History of the Telephone" (1910); and Alexander Graham Bell, "The Telephone" (1878). On the cable: Charles Bright, "The Story of the Atlantic Cable" (1903). For facts in the history of printing and descriptions of printing machines, see: Edmund G. Gress, "American Handbook of Printing" (1907); Robert Hoe, "A Short History of the Printing Press and of the Improvements in Printing Machinery" (1902); and Otto Schoenrich, "Biography of Ottmar Mergenthaler and History of the Linotype" (1898), written under Mr. Mergenthaler's direction. On the best-known New York newspapers, see: H. Hapgood and A. B. Maurice, "The Great Newspapers of the United States; the New York Newspapers," in "The Bookman", vols. XIV and XV (1902). On the typewriter, see Charles Edward Weller, "The Early History of the Typewriter" (1918). On the camera, Paul Lewis Anderson, "The Story of Photography" (1918) in "The Mentor", vol. vi, no. 19.; and on the motion picture, Colin N. Bennett, "The Handbook of Kinematography"; "The History, Theory and Practice of Motion Photography and Projection", London: "Kinematograph Weekly" (1911).

CHAPTER VII

For information on the subject of rubber and the life of Charles Goodyear, see: H. Wickham, "On the Plantation, Cultivation and Curing of Para Indian Rubber", London (1908); Francis Ernest Lloyd, "Guayule, a Rubber Plant of the Chihuahuan Desert", Washington (1911), Carnegie Institute publication no. 139; Charles Goodyear, "Gum Elastic and Its Varieties" (1853); James Parton, "Famous Americans of Recent Times" (1867); and "The Rubber Industry, Being the Official Report of the Proceedings of the International Rubber Congress" (London, 1911), edited by Joseph Torey and A. Staines Manders.

CHAPTER VIII

J. W. Roe, "English and American Tool Builders" (1916), and J. V. Woodworth, "American Tool Making and Interchangeable Manufacturing" (1911), give general accounts of great American mechanics.

For an account of John Stevens and Robert L. and E. A. Stevens, see George Iles, "Leading American Inventors" (1912); Dwight Goddard, "A Short Story of John Stevens and His Sons" in "Eminent Engineers" (1905), and R. H. Thurston, "The Messrs. Stevens, of Hoboken, as Engineers, Naval Architects and Philanthropists" (1874), "Journal of the Franklin Institute", October, 1874. For Whitney's contribution to machine shop methods, see Olmsted's "Memoir" already cited and Roe and Woodworth, already cited. For Blanchard, see Dwight Goddard, "A Short Story of Thomas Blanchard" in "Eminent Engineers" (1905), and for Samuel Colt, see his own "On the Application of Machinery to the Manufacture of Rotating Chambered-Breech Fire Arms, and Their Peculiarities" (1855), an excerpt from the "Minutes of Proceedings of the Institute of Civil Engineers", vol. XI (1853), and Henry Barnard, "Armsmear; the Home, the Arm, and the Armory of Samuel Colt" (1866).

CHAPTER IX

"The Story of Electricity" (1919) is a popular history edited by T. C. Martin and S. L. Coles. A more specialized account of electrical inventions may be found in George Bartlett Prescott's "The Speaking Telephone, Electric Light, and Other Recent Electrical Inventions" (1879).

For Joseph Henry's achievements, see his own "Contributions to Electricity and Galvanism" (1835-42) and "On the Application of the Principle of the Galvanic Multiplier to Electromagnetic Apparatus" (1831), and the accounts of others in Henry C. Cameron's "Reminiscences of Joseph Henry" and W. B. Taylor's "Historical Sketch of Henry's Contribution to the Electro-Magnetic Telegraph" (1879), Smithsonian Report, 1878.

"A List of References on the Life and Inventions of Thomas A. Edison" may be found in the Division of Bibliography, U. S. Library of Congress (1916). See also F. L. Dyer and T. C. Martin, "Edison; His Life and Inventions" (1910), and "Mr. Edison's Reminiscences of the First Central Station" in "The Electrical Review", vol. XXXVIII. On other special topics see: F. E. Leupp, "George Westinghouse, His Life and Achievements" (1918); Elihu Thomson, "Induction of Electric Currents and Induction Coils" (1891), "Journal of the Franklin Institute", August, 1891; and Alex Dow, "The Production of Electricity by Steam Power" (1917).

CHAPTER X

Charles C. Turner, "The Romance of Aeronautics" (1912); "The Curtiss Aviation Book", by Glenn H. Curtiss and Augustus Post (1912); Samuel Pierpont Langley and Charles M. Manly, "Langley Memoir on Mechanical Flight" (Smithsonian Institution, 1911); "Our Atlantic Attempt", by H. G. Hawker and K. Mackenzie Grieve (1919); "Flying the Atlantic in Sixteen Hours", by Sir Arthur Whitten Brown (1920); "Practical Aeronautics", by Charles B. Hayward, with an Introduction by Orville Wright (1912); "Aircraft; Its Development in War and Peace", by Evan J. David (1919). Accounts of the flights across the Atlantic are given in "The Aerial Year Book and Who's Who in the Air" (1920), and the story of NC4 is told in "The Flight Across the Atlantic", issued by the Department of Education, Curtiss Aeroplane and Motor Corporation (1919).